AN EXPERIMENTAL STUDY OF THE GENUS BIDENS
(ASTERACEAE) IN THE HAWAIIAN ISLANDS

AN EXPERIMENTAL STUDY
OF THE GENUS BIDENS (ASTERACEAE)
IN THE HAWAIIAN ISLANDS

BY

GEORGE W. GILLETT and ELVERA K. S. LIM

UNIVERSITY OF CALIFORNIA PRESS
BERKELEY AND LOS ANGELES
1970

University of California Publications in Botany
Advisory Editors: H. G. Baker, Peter Dixon, E. M. Gifford, F. T. Haxo,
M. E. Mathias, H. A. Mooney, C. H. Muller, P. C. Silva, Grady Webster

Volume 56

Approved for publication June 19, 1969
Issued January 20, 1970
Price, $2.00

University of California Press
Berkeley and Los Angeles
California

◇

University of California Press, Ltd.
London, England

CONTENTS

AN EXPERIMENTAL STUDY
OF THE GENUS BIDENS (ASTERACEAE)
IN THE HAWAIIAN ISLANDS

BY

GEORGE W. GILLETT and ELVERA K. S. LIM

INTRODUCTION

PURPOSE OF THE INVESTIGATION

ACCORDING TO THE revision of Sherff (1937) and several subsequent papers by him and other authors, the genus *Bidens* (Tribe Heliantheae, Subtribe Coreopsidineae) now totals approximately 280 species. Over two-thirds of these species occur in Africa and the New World, with about 100 species in each region. About sixty-five species occur in the Pacific where they are restricted to eastern Polynesia. The remaining species occur through the expanse of territory from Eurasia through the Malay Archipelago, New Guinea, and Australia. Of the sixty-five Polynesian species, over 40 are endemic to the extremely limited area (6,400 square miles) of the Hawaiian Islands. Of these species, over thirty are restricted to single islands and none has been found on all of the Hawaiian Islands.

The Hawaiian Islands are characterized by a relatively depauperate angiosperm flora, totaling only 216 (Fosberg, 1948) to 238 Balgooy, 1960) genera. This limited flora is related to the geographical isolation of Hawaii, it being more remote from the sources of colonizing species than any other archipelago. By contrast, the Fiji Islands (7,000 square miles), lying much closer to the major source region of New Guinea, have nearly twice the number present in Hawaii, some 445 angiosperm genera (Smith, 1951).

The remarkably high levels of speciation in Hawaii as portrayed by the taxonomists who contributed the data below clearly suggest that inquiries into the evolutionary dynamics of certain Hawaiian genera are very much in order.

Cyrtandra (Gesneriaceae)	159 species in Hawaii vs. 125 species in New Guinea
Pipturus Sect, Mamakea (Urticaceae)	13 species in Hawaii vs. 8 species, New Caledonia through Borneo to Formosa
Peperomia (Piperaceae)	48 species in Hawaii vs. 43 species in Africa
Wikstroemia Subg. Wikstroemia (Thymelaeaceae)	22 species in Hawaii vs. 27 species, Fiji through New Guinea and Japan
Bidens	Over 40 species in Hawaii vs. 29 species in South America

The most conservative speculation on the origin of the genus *Bidens* in the Hawaiian Islands is one that envisages an emigrant species arriving through most fortuitous circumstances, most likely as achenes attached to the feathers of migratory birds (Carlquist, 1966a, 1966b, 1966c). From this introduction, a singular event for an oceanic archipelago isolated from the nearest and most likely source region (North America) by over 2,000 miles of ocean, there evolved a multiplicity of expression that Man has interpreted as over forty species. Such a large number of species for a limited territory naturally poses some questions. Are

[1]

these well-marked species? Do the character differences that distinguish these species have a genetic basis? Are these species isolated by internal genetic barriers? What is the explanation for the natural intergrades? These are reasonable questions that merit serious attention by an experimental inquiry. The research reported here attempts to honor these questions and to enlighten our understanding of the processes of evolution as they operate in plants on oceanic archipelagos.

The Hawaiian species of *Bidens* occur on habitats that range from arid to semi-arid lava flows to dense rain forest, over moisture gradients extending from 0.5 meter to 5 meters of annual rainfall, and through elevations extending from sea level to over 2,000 meters. Few Hawaiian genera have the breadth of ecological diversity that is expressed in *Bidens*, and perhaps none have the remarkable capacity for vigorous growth and fertility in the environment of the greenhouse. The material studied in this work has thus provided an unusual opportunity for the investigation of evolutionary mechanisms. Species were chosen for the crossing program on the basis of distinctive morphological and ecological attributes that would in each case suggest a significant evolutionary divergence.

HISTORICAL SURVEY

The genus *Bidens* was described by Linnaeus (1753, 1754). The type species *B. tripartita* possesses achenial awns or aristae armed with retrorse barbs. Later authors varied in their opinion as to the circumscription of this genus, and some suggested alternate generic designations for the Polynesian species of the genus (table 1). The three names: *Bidens*, *Coreopsis*, and *Campylotheca*, have most often been applied to the Polynesian species by different authors.

Necker (1790) discarded the name *Bidens* on the basis that certain species possess more than two awns. The genus *Campylotheca* was proposed by Cassini (1827) because of the curved achenes of the Hawaiian species. This name was retained by many authors (table 1) even as recently as 1926 by Skottsberg, who later accepted the name *Bidens* (1935, 1944).

More than a century ago, Gaudichaud (1826) described the first Hawaiian *Bidens*, *B. micrantha*. Since then, other species were either recorded or described, especially in the work of Cassini (1827), Lessing (1832), DeCandolle (1836), Nuttall (1841), Schultz-Bipontinus (1856), Gray (1861), Hillebrand (1888), Skottsberg (1926, 1935, 1944), Degener (1933–1963), and Sherff (1937–1964). The most complete taxonomic treatment of this genus is by Sherff (1937), in which he included all of the Polynesian species, except one, in his Section Campylotheca. The Hawaiian species, *Bidens cosmoides*, was placed by itself in his second section, Degeneria, because of its extremely long and exserted styles.

ACKNOWLEDGMENTS

The financial assistance of the National Science Foundation (Grants GB 3336 and GB 7085) is acknowledged with appreciation. Many people have contributed invaluable assistance without which this study could not have been accomplished. It is a pleasure to acknowledge helpful discussions with Dr. Sherwin Carlquist of the Claremont Graduate Schools and Dr. A. C. Smith, University of Hawaii, on the origin and evolution of the Pacific biota. Our thanks are extended to Dr.

TABLE 1

THE GENERIC DISPOSITION BY VARIOUS AUTHORS OF THE SPECIES
OF BIDENS INDIGENOUS TO POLYNESIA

Author	Year	Bidens	Coreopsis	Camplyotheca	Adenolepsis	Delucia
Gaudichaud	1826	+				
Cassini	1827			+		+
Lessing	1832	+		+	+	
DeCandolle	1836	+	+	+		
Nuttall	1841	+				
Schultz-Bipontinus	1856	+				
Gray	1861	+	+	+		
Hillebrand	1888	+	+	+		
Drake del Castillo	1886–1892					
Skottsberg	1926	+		+		
	1935–1944	+				
Moore	1933	+				
Brown	1935	+				
Degener	1933–1963	+				
Sherff	1937–1964	+				

Roland Force, Director of the Bernice P. Bishop Museum, and to Dr. Pieter van Royen, chairman of the Botany Department of that institution, for providing herbarium resources and facilities that were critical to the successful completion of this study.

Valuable field assistance was accorded us by Edwin Bonsey, Donn Carlsmith, Norman Carlson, Jack Lee, James Lindsay, Noah Pekelo, William Sproat, and Dr. Frank Tabrah. Our work could not have proceeded without the timely assistance of these men. Thanks are extended to Dr. Yoneo Sagawa, Director of the Lyon Arboretum, University of Hawaii, who generously provided greenhouse space for experimental plants. The extensive collections of Drs. Otto Degener and Harold St. John were critical resources for the execution of this work. The skillful greenhouse assistance of Albert Gallo and Kenneth Nagata contributed much to the successful culture of experimental plants and our appreciation is extended to both men.

Voucher specimens of this research have been deposited in the herbaria of the Bernice P. Bishop Museum, Honolulu, and the University of California, Riverside.

MATERIALS AND METHODS

The following thirteen species were grown in the greenhouse and were included in the program of experimental crosses:

Bidens coartata Sherff   (*Gillett 1602*)
Bidens cosmoides (A. Gray) Sherff   (*Gillett 1763*)
Bidens ctenophylla Sherff   (*Gillett 1751, 1860*)
Bidens forbesii Sherff   (*Gillett 1765*)
Bidens fulvescens Sherff   (*Gillett 1810*)
Bidens hillebrandiana (Dr. del Cast.) Deg ex Sherff   (*Gillett 1772, 1773*)
Bidens macrocarpa (A. Gray) Sherff   (*Gillett 1603*)
Bidens mauiensis var. *mauiensis* (A. Gray) Sherff   (*Gillett 1872*)
Bidens mauiensis var. *cuneatoides* Sherff   (*Gillett 1873*)
Bidens menziesii var. *menziesii* (A. Gray) Sherff   (*Gillett 1802*)
Bidens menziesii var. *filiformis* Sherff   (*Gillett 1745*)
Bidens menziesii var. *filiformis*   (*Gillett 1756*)
Bidens molokaiensis (Hilleb.) Sherff   (*Gillett 1807*)
Bidens skottsbergii Sherff   (*Gillett 1753, 1858*)
Bidens wiebkei Sherff   (*Gillett 1819*)
Bidens (*B. forbesii* × *B. menziesii*) Putative Hybrid   (*Gillett 1888*)

These collections were identified from herbarium specimens at the Bernice P. Bishop Museum, following Sherff's taxonomic treatment of the genus.

Buds collected from field and greenhouse plants were fixed in four parts chloroform to three parts absolute alcohol to one part glacial acetic acid. Aceto-carmine smears were used and slides were made permanent by running through the following series of solutions:

1. 45 percent glacial acetic acid: 95 percent alcohol (1:1);
2. 95 percent alcohol: 100 percent TBA (tertiary butyl alcohol) (1:1);
3. 100 percent TBA;
4. 100 percent TBA: 100 percent xylene (1:1);
5. 100 percent xylene.

For thin-layer chromatography (Stahl, 1965), leaves were dried and stored in a dark cabinet. For each chromatogram, 0.1 gram of dried leaves was ground and extracted first in 10 ml of boiling distilled water for five minutes then three times with 10 ml ethyl acetate and concentrated to 1 ml. The ethyl acetate extract was concentrated in an atmosphere of nitrogen gas to minimize oxidation, then was applied as a small spot to the chromatoplate, using polyamide (Merck) as an adsorbent. Thin-layer (.25 mm) plates were made by mixing thoroughly 15 g of polyamide with 70 ml of methanol, and the homogenized slurry applied to five 20 cm × 20 cm plates with a Desaga-Brinkman applicator. Solvents were left in the tank for not less than one half hour to balance the atmosphere before the plates were run. The tank was shaken before use. The plates were run two-dimensionally. The following solvent system was employed: (1) (vertical axis of chromatogram), a mixture of ethyl acetate and methanol (4:1); (2) (horizontal axis of chromatogram), methanol. Dried plates were inspected under a mineralight USVL-13 ultraviolet lamp and the positions of spots were recorded.

Reciprocal crosses were made between species whenever possible. Applications of pollen were made either by rubbing a detached floret on the exserted stigma, or by transferring the pollen with a fine needle. Pollinated capitula were then tagged and covered with organdy bags.

Comparisons of the hybrids and parents were based on the following characters: habit, branching pattern, leaf form, inflorescence, and achene structure.

Drawings of the branching pattern were made from greenhouse plants, taking average measurements. Leaves were traced at life size. Drawings of achenes were obtained by tracing mature structures on a Reichert Visipan enlarger. The outlines of the capitula were done by direct tracing from photographic prints. The diameters of the capitula were measured across the involucral bracts.

DISTRIBUTION, MORPHOLOGY, AND ECOLOGY OF SPECIES

Although the species of *Bidens* have been described in Sherff's monograph (1937), it is necessary to discuss the more important morphological and ecological features of the species included in this study.

Bidens coartata Sherff

This species occurs on the Koolau Range of the island of Oahu (fig. 1, *h*). Greenhouse material (*Gillett 1602*) was grown from achenes collected on the Mt. Olympus Trail, southeast end of the Koolau Range, at an elevation of 530 m in *Metrosideros* forest. The species grows in association with *Metrosideros, Hibiscus, Cyanea, Clermontia, Cheirodendron,* and other indigenous rain forest genera. Wild plants grow to about 0.9 m high, while the greenhouse plants (fig 2, *a*) reach a height of about 0.8 m. The lower branches arise at angles of about 60 degrees with the main stem and each terminates with a loosely corymbose inflorescence. The cauline leaves are generally 3- to 5-parted, with ovate-lanceolate segments. These segments have acuminate apices and oblique bases; the margins are serrate with broad teeth. The young branches, inflorescences, and upper and lower leaf surfaces are pubescent with uniseriate hairs. The capitula are about 2.3 cm in

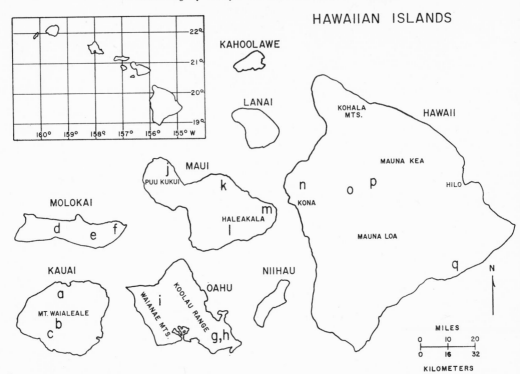

Fig. 1. Distribution of Hawaiian *Bidens* included in this study. *a, B. forbesii (Gillett 1765)*; *b, B. cosmoides (Gillett 1763)*; *c, B. forbesii* × *B. menziesii* putative hybrid *(Gillett 1888)*; *d, B. molokaiensis (Gillett 1807)*; *e, B. menziesii* var. *menziesii (Gillett 1802)*; *f, B. wiebkei (Gillett 1819)*; *g, B. macrocarpa (Gillett 1603)*; *h, B. coartata (Gillett 1602)*; *i, B. fulvescens (Gillett 1810)*; *j, B. mauiensis* var. *cuneatoides (Gillett 1873)*; *k, B. hillebrandiana (Gillett 1773)*; *l, B. mauiensis* var. *mauiensis (Gillett 1872)*; *m, B. hillebrandiana (Gillett 1772)*; *n, B. ctenophylla (Gillett 1751, 1860)*; *o, B. menziesii* var. *filiformis (Gillett 1745)*; *p, B. menziessi* var. *filiformis (Gillett 1756)*; *q, B. skottsbergii (Gillett 1753)*.

diameter at anthesis, each consisting of 5 ray florets and 10–22 disc florets. The achenes are black and are from 7 to 10 mm long and 1 mm broad. The achene body and apex are sparsely ciliate, and more so on the lateral margins. Short, usually antrorsely barbed awns arise subapically, or the awns may be absent (fig. 23).

Bidens cosmoides (Gray) Sherff

This species is restricted to the island of Kauai in the dense rain forest of the Kokee area (fig. 1, *b*). The greenhouse material was grown from young plants collected near Halemanu, Kokee, at an elevation of 1,100 m *(Gillett 1763)*. The species grows in dense vegetation, in deep shade under *Acacia koa, Metrosideros, Cyanea leptostegia,* and other rain forest species. In the natural habitat, the plants are shrubby and sprawling with long reclining branches rooting at the nodes and grow to about 2 m high. Greenhouse plants have much shorter branches and grow to about 0.4 m tall. The cauline leaves of the wild plants are very large and thin, measuring up to 20 cm in length including the petioles. They are ternately or pinnately 3- to 9-parted (fig. 3, *j*), but a few simple leaves intermingle at different positions on the same branch. The greenhouse plants have the leaves very much reduced and lighter in color, but retain the wild leaf form. The leaflets are pubescent with uniseriate hairs on the upper and lower surfaces and are lanceolate to obovate-lanceolate, with acuminate apices and serrate margins that are sparsely ciliate. The capitula are solitary on decurved peduncles, and (including rays) are up to 7.5 cm in diameter at anthesis. Each capitulum consists of 7–14 ray florets and 20–50 disc florets. The styles are unusually long, up to 2.5 cm, and are about 2 cm longer than the dark brown staminal tubes. The styles and anther apices are long acuminate.

The cultivated plants did not do well under greenhouse conditions and failed to flower even after a period of two years. Pollinations were made from capitula collected in the field and brought to the greenhouse.

Apparently the mature achenes of *Bidens cosmoides* had escaped other collectors' notice by the peculiar way they are enveloped by a receptacular bract. They are dark brown, up to 1.0 cm long, and about 2 mm broad. Each is clasped by an abaxial receptacular bract that encloses the entire structure. The achenes are contorted longitudinally into the configuration of a question mark, but they are not coiled or twisted (fig. 24).

This species has been considered "a strange and anomalous species" by Sherff who placed it alone in Section Degeneria. This is based mainly on its long styles and unusually large capitula, certainly the largest in Hawaiian *Bidens*. Morphologically, this separation is well supported by the peculiar envelopment of the mature achene by a receptacular bract, a feature that is not found in other species of *Bidens*. This condition, obviously a product of Hawaiian *Bidens*, is falsely suggestive of a unique monotypic genus in the subtribe Madiinae.

Bidens ctenophylla Sherff

This species (fig. 4, *c*) occurs on the island of Hawaii, on lava flows in an area extending from Huehue to Puu Waawaa, between Kona and Kamuela (fig. 1, *n*). The greenhouse material was grown from achenes and seedlings collected from Huehue Ranch, Hawaii, on an extremely arid, hot, and weed covered "A-A" lava flow, at about 150 m (*Gillett 1751, 1860*). It occurs over quite a large area, scattered among *Diospyros, Capparis, Reynoldsia,* and other dry-forest plants. In the field it grows to about 2 m high, but the greenhouse plants attain a height of about 1.1 m. The main stem is woody, the lower branches arise from it at angles of about 30 degrees, but later tend to bend upright until nearly level with the upper branches. The species is easily recognized by its simple cauline leaves that are ovate or rhomboid-ovate with long acuminate apices and cuneate bases (fig. 3, *d*). The leaf margins are irregularly serrate. The upper leaf surface is glabrous, the lower surface sparsely ciliate. The inflorescences are compact, corymbose panicles, consisting of numerous small capitula. The capitula are of about 1.8 cm in diameter at anthesis, consisting of 5–7 ray florets and 11–18 disc florets. Sometimes there are 1–2 tubular ray florets present in a capitulum. The achenes are 1.1–1.2 cm long, and 0.1 cm broad, dark brown, straight, or slightly twisted, and

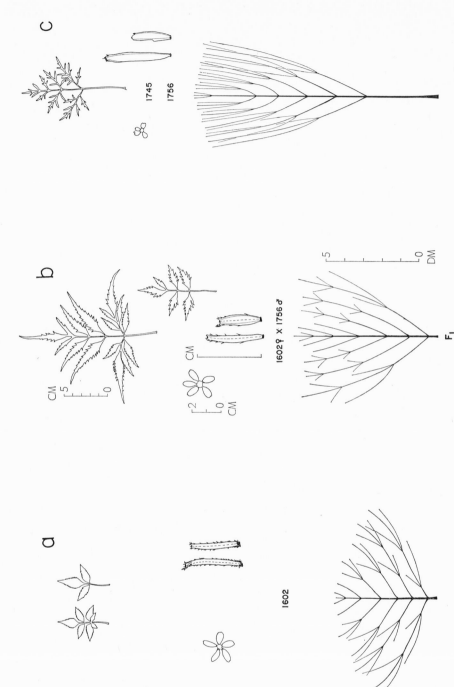

Fig. 2. *a*, *Bidens coartata*; *b*, experimental F₁ hybrid, *B. coartata × B. menziesii* var. *filiformis*; *c*, *B. menziesii* var. *filiformis*.

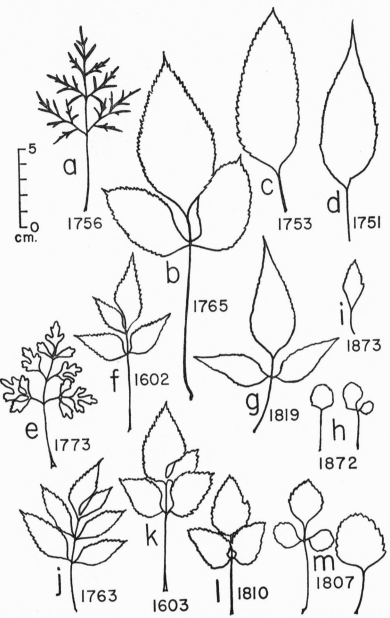

Fig. 3. Leaf form in Hawaiian species of *Bidens*. Drawings traced from photographs. *a, B. menziesii* var. *filiformis; b, B. forbesii; c, B. skottsbergii; d, B. ctenophylla; e, B. hillebrandiana; f, B. coartata; g, B. wiebkei; h, B. mauiensis* var. *mauiensis; i, B. mauiensis* var. *cuneatoides; j, B. cosmoides; k, B. macrocarpa; l, B. fulvescens; m, B. molokaiensis.*

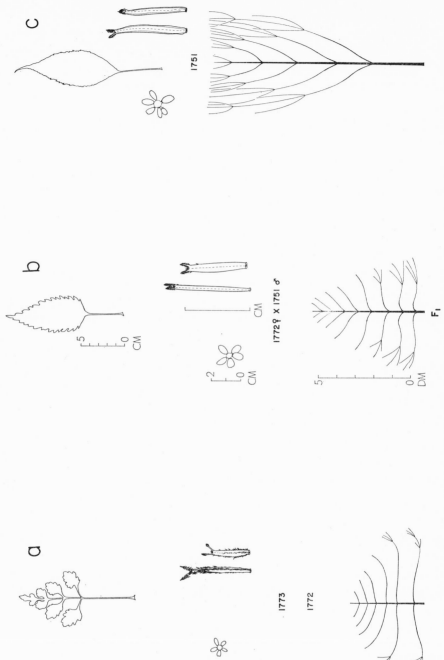

Fig. 4. *a*, *Bidens hillebrandiana*; *b*, experimental F₁ hybrid, *B. hillebrandiana × B. ctenophylla*; *c*, *B. ctenophylla*.

slender; the body is somewhat smooth except at the apex where it is crowned by a few hairs, and two (or rarely one) retrorsely barbed apical awns, which are generally not divergent. In general, the morphological characters of the wild plants are exhibited by the greenhouse plants.

Bidens forbesii Sherff

This species occurs on the island of Kauai near sea level, along the north coast (fig. 1, *a*). The greenhouse material was grown from achenes and seedlings collected from the Waioli Valley near Hanalei, about 200 yards from the seashore (*Gillett 1765*). Here the species grows on a steep, well-shaded road bank. In nature, it grows to a height of about 2 m, but the greenhouse plants grew to about 0.7 m. The main stem is woody and vertical with lower branches arising almost horizontally from it; each branch eventually bends upright, with a loosely corymbose inflorescence. The cauline leaves are simple to trifoliate with ovate-lanceolate leaflets (fig. 3, *b*). The leaflets have long-acuminate apices and usually nearly sessile oblique bases; the margins are closely serrate with long mucronate-inflexed teeth. The capitula are about 2.5 cm in diameter at anthesis and each consists of 5–8 ray florets and 28–32 disc florets. The achenes are black and slender, slightly twisted, and are up to 1.1 cm long and 0.1 cm broad. The achene body and lateral margins are sparsely covered with hairs, there are two apical, retrorsely barbed awns.

Bidens hillebrandiana (Dr. del Cast.) Deg. ex Sherff

This species (fig. 4, *a*) occurs on the windward side of the islands of Molokai, Maui, and Hawaii on or near the seacoast where it is often exposed to sea spray. The achenes and seedlings were collected from Hana and Maliko Bay on the island of Maui. The species is quite distinctive in its herbaceous and prostrate to erect habit, also in its branching pattern, dissected leaf form, and especially in its strongly awned achene structure. The greenhouse plants tend to be erect. The species grows to about 0.5 m high in nature, but the greenhouse plants are approximately 0.4 m tall. The lower branches are horizontal, somewhat reclining at right angles to the main stem; the upper branches are at angles less than 90 degrees to the main stem. The branches end in spreading and loose inflorescences. The cauline leaves are mainly bipinnately divided into obovate or oblong segments with lobed margins (fig. 3, *e*); a form that does not occur in other Hawaiian *Bidens*. The capitula are about 1.4 cm in diameter at anthesis, each consisting of 5–6 ray florets and 11–21 disc florets. The awned achenes are distinctive in that they possess four of the five features that Carlquist (1966*b*) cites as efficient dispersal mechanisms among Hawaiian *Bidens:* spreading awns, retrorsely barbed awns, the presence of upward-pointing hairs on the lateral margins, and the presence of upward-pointing hairs on the dorsiventral margins of the achene body. The dorsiventral margins form ridges from near the apices, and from these short awns sometimes arise; often giving the achene as many as five awns. The achenes are dark grayish brown, with lateral and dorsiventral margins. They are up to about 0.6–1 cm long and 0.1 cm broad (fig. 23, *c, f*).

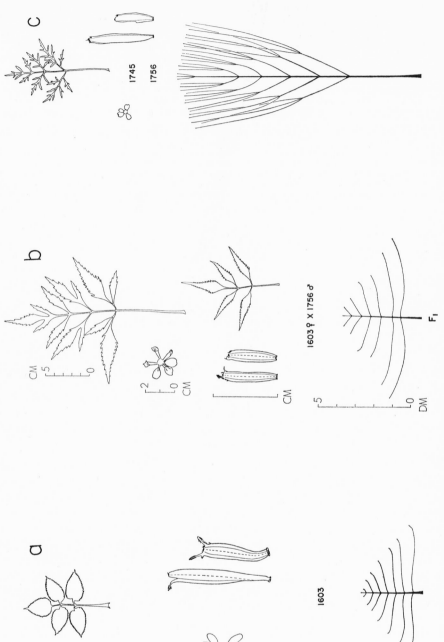

Fig. 5. *a*, *Bidens macrocarpa*; *b*, experimental F₁ hybrid, *B. macrocarpa* × *B. menziesii* var. *filiformis*; *c*, *menziesii* var. *filiformis*.

Hana race.—The greenhouse material for this race was grown from seedlings collected from Hana, East Maui (fig. 1, *m*), on red cinders of the headland adjacent to the Hana wharf, growing among *Osteomeles* and weeds at sea level (*Gillett 1772*).

Maliko Bay race.—The greenhouse plants for this race were grown from achenes and seedlings collected from Maliko Bay, north coast of East Maui (fig. 1, *k*). The species occurs on the seaward side of high clay bluffs on Maliko Bay at an elevation of about 25 to 34 m (*Gillett 1773*). It grows on patches of exposed soil in association with assorted weeds, especially *Cynodon, Cenchrus, Leucaena,* and others. The greenhouse and wild plants of both races do not show any conspicuous differences in morphology and size.

Bidens macrocarpa (Gray) Sherff

This species (fig. 5, *a*) occurs on Oahu, on both the Waianae and Koolau ranges. The greenhouse material was grown from achenes collected from the Mt. Olympus Trail, southern Koolau Range, at an elevation of about 650 m (*Gillett 1603*) (fig. 1, *g*). This race grows in wet *Metrosideros* forest in association with *Clermontia, Scaevola, Pelea,* and *Pittosporum.* In the field, it grows up to 1.4 m high, but the greenhouse plants grow to a height of about 1.0 m. It is shrubby with the lower part of the main stem woody, stout, and tetragonal with very short internodes; with the shortest internode measured about 3 mm long. The lower branches are almost horizontal, arising at about right angles to the main stem. The upper branches are at more acute angles with the main stem. The branches end in loose corymbose inflorescences. The cauline leaves are mainly 3- to 5-parted with ovate to ovate-lanceolate segments, each with a cuspidate apex and an oblique base (fig. 3, *k*). Leaf margins are closely and sharply serrate with inflexed teeth. The capitula, quite showy, are up to 3.5 cm in diameter at anthesis, each consisting of 5–7 ray florets and 11–26 disc florets. The achenes are the largest among Hawaiian species, being 1.1 to 1.8 cm long and 2.5 mm broad (fig. 5, *a*). They are light brown, slightly twisted, and with lateral marginal wings often extending upward to form two short teeth just below the two sparsely barbed subapical awns. The awns are rather irregular in length and are rarely straight.

Bidens mauiensis (Gray) Sherff

This species (fig 6, *c*) occurs on the island of Maui near the seacoast, slightly above sea level on extremely arid, hot, sandy, or rocky hills. The greenhouse material was grown from seedlings collected from Manawainui Gulch and from the Waihee sand bluffs of the island of Maui. The species is easily recognized by its thickened leaves, large scapose capitula, and herbaceous, procumbent habit. These characters are suggestive of *Bidens molokaiensis.* The species differs especially in its laterally winged achenes that are distinctive among all the Hawaiian species of *Bidens.*

Manawainui Gulch race (var. *mauiensis*).—The greenhouse material for this race was grown from seedlings collected from Manawainui Gulch, south coast of east Maui (fig. 1, *l*) at about 60 m above the sea (*Gillett 1872*). This race grows on extremely hot, arid, and exposed rocky slopes. It is associated with indigenous *Euphorbia* and dry-land weeds. It is procumbent in habit with the branches lying

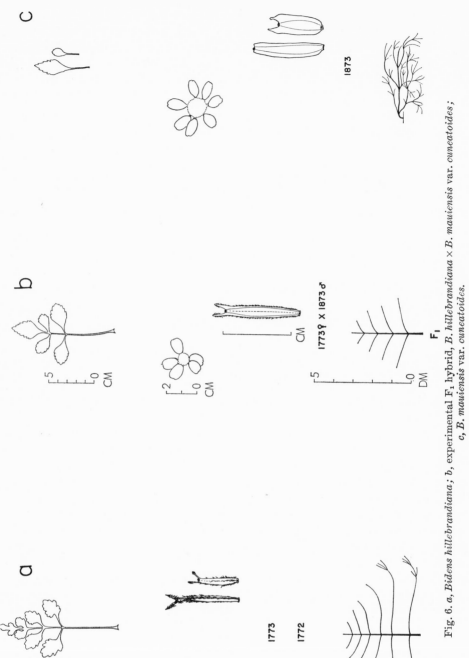

Fig. 6. *a*, *Bidens hillebrandiana*; *b*, experimental F₁ hybrid, *B. hillebrandiana* × *B. mauiensis* var. *cuneatoides*; *c*, *B. mauiensis* var. *cuneatoides*.

on the soil surface. The solitary capitula are held erect by very long peduncles. This habit is retained by the greenhouse plants that form a compact mass covering the soil surface in the pots. The cauline leaves are thick and fleshy and are mainly simple, but may be 2- to 3-parted; blades are mainly rhomboid-ovate with serrate margins (fig. 3, *h*). The capitula are showy, about 3.8 cm in diameter at anthesis, consisting of 6–11 ray florets and 18–42 disc florets. The achenes are about 0.7–1 cm long and 2.5–3 mm broad, light brown, and the broad marginal wings expand apically to form 2 stubby awnlike structures.

Waihee race (var. *cuneatoides*).—The greenhouse material for this race (fig. 6, *c*) was grown from achenes and seedlings collected from Waihee, north coast of west Maui (fig. 1, *j*), growing on exposed, hot, sandy bluffs about 20 m above sea level (*Gillett 1873*). This race is in association with *Lipochaeta, Scaevola coriacea, Nama sandwicensis, Fimbristylis,* and other species that usually occur on beaches. It differs from the Manawainui Gulch race in its leaf form (fig. 3, *i*) which is mainly simple and oblong, with an acuminate apex and serrate margins. The leaf blades are thinner than those of the foregoing race. It is closely similar to the Manawainui Gulch race in its habit and other morphological characters, especially in the achene structure (fig. 23, *d, g*).

Bidens menziesii (Gray) Sherff

This species is one of the most widely distributed of Hawaiian *Bidens,* occurring on the islands of Molokai, Maui, and Hawaii. The species grows on relatively arid, sunny and windswept slopes, cliffs, and plateaus, sometimes on "A-A" lava flows, at elevations of 450 to 2,000 m. It is most easily recognized by its finely dissected leaves.

HAWAII

Saddle Road race (var. *filiformis*).—The greenhouse material for this race (fig. 2, *c*) was grown from achenes and seedlings collected on the central plateau (fig. 1, *p*), near the Saddle Road, at an elevation of about 1,800 m (*Gillett 1756*). This race grows on arid, windblown, fine soil in the cracks of lava blocks. It grows in association with *Dodonaea, Osteomeles,* and *Dubautia.* The plants are up to more than 3 m tall in the field. They are shrubby with a virgate main stem, the lower branches arising at angles less than 30 degrees with the main stem. The branches terminate in compact, corymbose paniculate inflorescences. The leaves are pinnately to bipinnately dissected into filiform segments, with the segment margins closely ciliate. The corymbose inflorescence consists of numerous small capitula. The capitula are about 1.5 cm in diameter, each consisting of 4–6 ray florets and 5–16 disc florets. The brown achenes are 0.6–1.2 cm long and 1.2–1.5 mm broad, straight or slightly bent and with lateral marginal wings (fig. 23, *e*). One or more toothlike structures usually arise from the wings. The achene body is smooth with the apex crowned by uniseriate hairs. Terminal awns are sometimes weakly developed.

Puu Waawaa race (var. *filiformis*).—The greenhouse material for this race was grown from achenes and seedlings collected on Puu Waawaa Ranch (fig. 1, *o*), at an elevatioon of 1,100 m (*Gillett 1745*). This race grows in undisturbed native vegetation, in association with *Osteomeles, Dodonaea, Coprosma, Metrosideros,* and *Sophora*. Its habit and other morphological characters are closely similar to that of the Saddle Road race (fig. 2, *c*). Greenhouse plants show slight achene differences from plants of the Saddle Road race (fig. 23, *c, e*).

MOLOKAI

Molokai Ranch race (var. *menziesii*).—The greenhouse material for this race was grown from seedlings collected on the south slope of Molokai, above Kaunakakai (fig. 1, *e*), at an elevation of about 812 m, on badly eroded, deeply gullied soil (*Gillett 1802*). It is in association with *Grevillea* and *Dodonaea* and is more easily recognized by its unusually large, finely dissected leaves (fig. 8, *b*). Wild plants grow to a height of 2–3 m. The main stem is virgate and is unbranched on its lower part. This feature is also noted in the greenhouse plants. The leaves of this race are the largest among the Hawaiian species, and may be as long as 40 cm or more. The achenes are more slender and long and may have short awns (fig. 23, *b*).

Bidens molokaiensis (Hillebr.) Sherff

This species (fig. 7, *a*) occurs on the north coast of the island of Molokai, on the crests of the cliffs (fig. 1, *d*). The greenhouse material was grown from achenes and seedlings collected from near Hoolehua, at an elevation of about 150 m (*Gillett 1807*), amid an old trash dump, on windswept grassland exposed to the ocean spray during stormy weather. The species was collected from *Cynodon dactylis* sod. It is a low, diffuse, procumbent herb with slender branches that extend to more than 1 m long. Greenhouse plants exhibit this habit. The leaves are usually simple, but occasionally there are 2- to rarely 3-parted leaves (fig. 3, *m*). They are rhomboid-ovate to somewhat deltoid, with subobtuse to subacuminate apices and subcordate bases. The leaf margins are coarsely crenate. The capitula are borne on upright peduncles, are about 3.0 cm in diameter at anthesis, each consisting of 6–11 ray florets and 18–42 disc florets. The achenes are black, 0.7–1.2 cm long and 1 to 1.5 mm broad and slightly winged on the lateral, more or less ciliate, margins. Terminal awns are sometimes present, and while they are usually short with retrorse barbs, they attach the achenes to the surface of cotton cloth.

The habit and leaf form of *Bidens molokaiensis* are suggestive of *B. mauiensis*, but the two species are very different in their achene structure and ecology.

Bidens skottsbergii Sherff

This species (fig. 7, *c*) occurs on the south coast of the island of Hawaii. The greenhouse material was grown from seedlings collected on the 1750 "A-A" lava flows near Black Sand Beach (fig. 1, *q*), at an elevation of about 12 m (*Gillett 1753*). In the above habitat the species occurs with *Metrosideros, Scaevola, Pluchea,* and *Polypodium,* and the plants grow to nearly 2 m high. This species can be recognized by its simple leaves, which are glabrous and shiny with rounded to cuneate bases, and by the strongly awned achenes. The main stem is woody with lower branches arising from it at almost right angles but these later bend upright and bear terminal, loosely paniculate inflorescences nearly level with the inflorescences of the upper branches. The leaves are oblong-ovate, with subobtuse apices. The leaf margins are serrate with broad teeth. The capitula are about 3.0 cm in diam-

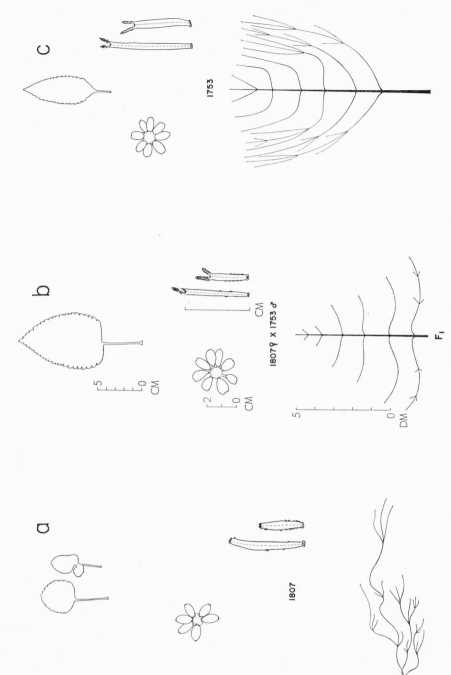

Fig. 7. *a*, *Bidens molokaiensis*; *b*, experimental F₁ hybrid, *B. molokaiensis* × *B. skottsbergii*; *c*, *B. skottsbergii*.

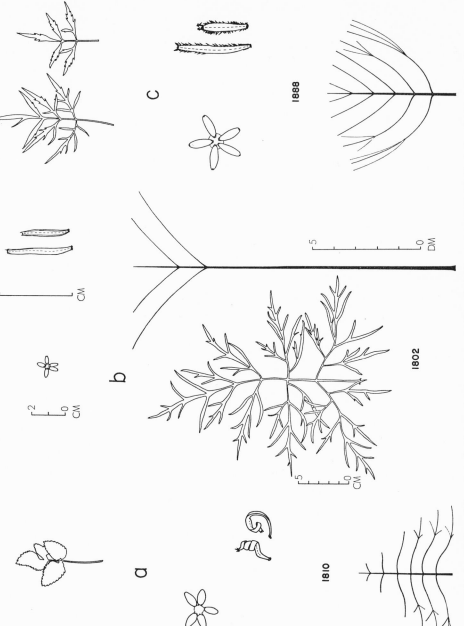

Fig. 8. *a*, *Bidens fulvescens*; *b*, *B. menziesii* var. *menziesii*; *c*, Putative natural hybrid, *B. forbesii* × *B. menziesii*.

eter at anthesis. Each capitulum consists of 8–10 ray florets and 26–32 disc florets. The achenes are black, 0.9–1.3 cm long and 1–1.5 mm broad, with two retrorsely barbed apical awns that are 1–3 mm long and often are not divergent (fig. 23, *a*). The achene body is smooth.

Bidens fulvescens Sherff

This species (fig. 8, *a*) occurs on the island of Oahu, on the Waianae Range. The greenhouse material was grown from achenes collected from Mauna Kapu Ridge (fig. 1, *i*), on windswept, moderately dry slopes, at an elevation of about 820 m (*Gillett 1810*). This species grows up to a height of about 1.5 m in nature. The main stem is woody at the base, with lower branches arising almost horizontally from it. These lower reclining branches often root at the nodes where they touch the soil surface. The cauline leaves are usually 3- to 5-parted with ovate-lanceolate segments. Leaf margins are closely and irregularly dentate with scattered uniseriate hairs between the teeth. The branches are terminated by corymbose inflorescences with capitula about 2.5 cm in diameter. Each capitulum consists of 5–8 ray florets and 14–22 disc florets. This species is one of several Hawaiian species of *Bidens* that possess coiled, awnless achenes. The achenes are black and smooth and are up to 1 cm long, and 0.1 cm broad, and bear no evidence of any provision for dispersal.

Bidens forbesii × B. menziesii Putative Hybrid (*Gillett 1888*)

This population, not identifiable with any of the current species, occurs near Waimea Canyon on the Kokee highway at about 1,000 m in an open road cut. Associated species include an overstory of *Acacia koa* and *Metrosideros* and an understory of assorted weeds. This area (fig. 1, *c*) is an ecotone between the dense rain forest of the highlands of Kauai and the arid lowlands of the southwest coast.

Plants of this population (fig. 8, *c*) grow to approximately 1 m tall and have a spreading habit with the lower branches at about 60 degrees from the central axis. The inflorescences are spreading on the upper branches, with many small capitula ca. 2 cm in diameter. The achenes are straight, dark brown, 0.8 to 1.2 cm long, bearing retrorsely barbed, short apical awns. A tenable interpretation would be that the population, in itself highly variable in leaf morphology, is derived from crosses between *B. forbesii* and *B. menziesii*.

Bidens wiebkei Sherff

This species (fig. 9, *a*) occurs on the island of Molokai. The greenhouse material was grown from achenes and seedlings collected (*Gillett 1819*) from the top of a high bluff near Halawa Bay, East Molokai (fig. 1, *f*). It grows on a steep north-facing slope at an elevation of about 270 m, in a dense mat of the grass *Melinus*, in association with young *Metrosideros, Wikstroemia, Cocculus*, and *Sphenomeris*. Both the wild and greenhouse plants grow to a height of about 0.8 m. The main stem is woody with lower branches arising at angles approaching 90 degrees with it. These branches end in compact corymbose inflorescences that consist of numerous capitula. The cauline leaves are simple to 3- to 5-parted. The lanceolate segments have acuminate apices and sharply serrate margins (fig. 3, *g*). The

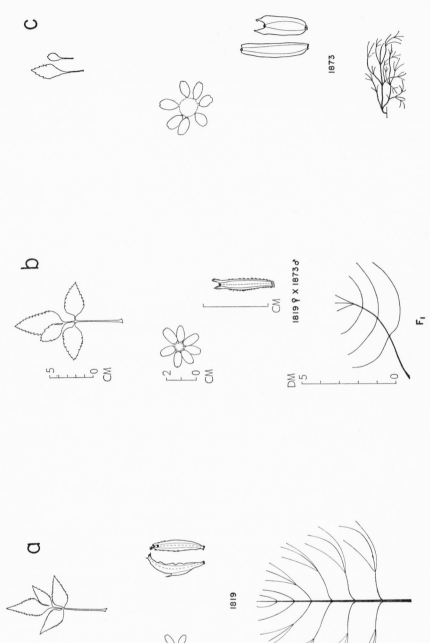

Fig. 9. *a, Bidens wiebkei; b,* experimental F₁ hybrid, *B. wiebkei* × *B. mauiensis* var. *cuneatoides*; *c, B. mauiensis* var. *cuneatoides.*

capitula are up to about 2.3 cm in diameter at anthesis, each consisting of 4–6 ray florets and 13–16 disc florets. The ray florets are not fully spread and the disc florets are compact. The anthers of the disc flowers of the capitula of certain plants do not rise above the corolla tubes, thus making the stylar branches appear more conspicuous and giving the capitula a very different appearance from those of other Hawaiian species. The achenes are 0.7–0.9 cm long and 1.5–2.0 cm broad, are slightly twisted, and with wavy marginal wings (fig. 23, *b*, *d*, *f*, and *h*). The achenes are strongly to weakly awned.

RESULTS

SUMMARY OF THE MORPHOLOGY OF GREENHOUSE PLANTS

1. Habit.—The habit and the branching pattern within a single species remain relatively constant under greenhouse conditions. The habits of the greenhouse plants fall into four types:
   a. Low, diffuse, procumbent herb. *Bidens mauiensis* and *B. molokaiensis* are of this type. Their lateral branches grow close to the soil surface with solitary inflorescences standing upright on long scapes.
   b. Low erect herb. This habit is characterized by *Bidens hillebrandiana,* both in the natural habitat and in the greenhouse. Injury to wild plants sometimes produces a decumbent habit, not an uncommon expression in habitats subjected to the northeast trade winds.
   c. Low woody shrub. This type includes *Bidens coartata, B. macrocarpa, B. wiebkei, B. forbesii, B. fulvescens,* and *B. cosmoides.* In all of these species the lower lateral branches generally extend almost horizontally from quite near the soil surface.
   d. Tall woody shrub. This includes *Bidens menziesii, B. ctenophylla,* and *B. skottsbergii.* The lower part of the main stem is generally thick and woody, up to 4–5 inches in diameter, without any lateral branches.
2. Branching Pattern.—The branching patterns of all thirteen species are of four types:
   a. Slender branches at angles less than 40 degrees with the main stem, trailing and with irregular lengths. This pattern is seen in the procumbent species *Bidens mauiensis* and *B. molokaiensis.*
   b. Branches at right angles to the erect main stem, extending horizontally. *Bidens hillebrandiana, B. macrocarpa, B. fulvescens,* and *B. cosmoides* are of this type
   c. Branches at angles slightly less than 90 degrees to the main stem but uniformly curved, turning upright at ends. This type is seen in *Bidens wiebkei* and *B. skottsbergii.*
   d. Branches at angles less than 90 degrees with the main stem and remaining nearly straight. This occurs in *Bidens coartata, B. ctenophylla, B. menziesii,* and *B. forbesii.*

   In *Bidens menziesii, B. skottsbergii, B. forbesii, B. wiebkei,* and *B. ctenophylla,* the tops of the branches come to about the same level. Rooting from the nodes of the lateral branches that touch soil has been observed both in field and greenhouse plants of *B. cosmoides* and *B. fulvescens.*

3. Leaf form (fig. 3).—The cauline leaf form in these species (representative of all Hawaiian *Bidens*) ranges from simple and entire to finely dissected with a series of intermediate forms. Both of the extreme forms are found among species from hot, arid, and exposed areas. *Bidens mauiensis, B. molokaiensis, B. ctenophylla,* and *B. skottsbergii* have mainly simple cauline leaves. The leaves of *Bidens mauiensis* are fleshy, as are those on field plants of *B. molokaiensis. Bidens menziesii* has finely bipinnately dissected leaves with filiform segments. *Bidens hillebrandiana* also has bipinnately dissected leaves but with broader segments than those of *B. menziesii.* Species from rain forest areas usually have pinnately divided leaves with broad segments: *Bidens coartata, B. macrocarpa, B. fulvescens, B. forbesii,* and *B. cosmoides.*

4. Inflorescence.—The inflorescence falls into three types: solitary, loosely corymbose with many capitula, and compactly corymbose with many capitula. *Bidens mauiensis, B. molokaiensis,* and *B. cosmoides* fall into the first type, the inflorescences in each case having long peduncles. *Bidens cosmoides* has a solitary inflorescence with decurved peduncles and pendulous capitula. The inflorescences of *Bidens skottsbergii, B. fulvescens,* and *B. wiebkei* are loosely corymbose, with fewer capitula. *Bidens menziesii* and *B. ctenophylla* have compactly corymbose inflorescences (much more so in the latter) with numerous capitula.

5. Achene structure (figs. 23, 24).—Achene structure in these species is very diverse: awned to awnless; a well-barbed body to a smooth body; laterally winged to wingless; straight body to coiled body. Achenes well equipped for dispersal as well as achenes poorly equipped for dispersal are produced by species from hot, arid, and exposed habitats of coastal and upland regions. Rain forest species, however, have achenes that usually are poorly equipped for dispersal. The achenes of *Bidens hillebrandiana,* a coastal species, are perhaps most efficiently equipped for dispersal; they have long, divergent and retrorsely barbed awns and the bodies are closely covered with barbs. The achenes of the coastal *Bidens skottsbergii* have long, parallel to divergent, barbed awns but the achene bodies are smooth. The achenes of *Bidens ctenophylla* are similar to those of *B. skottsbergii* but the awns are often shorter and less divergent. The achenes of *Bidens mauiensis,* a coastal species, are awnless with broad marginal wings and smooth bodies, with apparently a minimal adaptation to dispersal. The achenes of *Bidens menziesii* and *B. molokaiensis* are also often awnless (or with vestigial awns) with somewhat smooth bodies. The achenes of *Bidens wiebkei* have vestigial or well-developed awns but the achene bodies are slightly twisted and with undulate wings. The achenes of the other species are of intermediate structure except in *Bidens macrocarpa, B. fulvescens,* and *B. cosmoides.* The achenes of *Bidens macrocarpa* have vestigial awns, smooth bodies, and are the largest in the Hawaiian species of *Bidens,* while those of *B. fulvescens* are coiled. The achene bodies of *Bidens cosmoides* are bent longitudinally but are not coiled, and each achene is clasped by a receptacular bract that envelops nearly the entire achene, reflecting the diagnostic character of the subtribe Madineae.

Cytology of the Species

Previous counts for this genus indicate that it has the following basic numbers of chromosomes; $x = 10$, 11, 12, 13, 14, and 17 (Löve and Solbrig, 1964; Cave, 1956–1963; Ornduff, 1967, 1968). The lowest chromosome number recorded for this genus is $n = 10$ for *Bidens angustissima* var. *linifolia* (Turner et al., 1961), and the highest is $n = 38$, recorded for *B. pilosa*. The somatic number of two Hawaiian species, *Bidens conjuncta* and *B. sandwicensis*, was reported by Skottsberg (1953) to be 70 and 72, respectively. The chromosome number of other Hawaiian species had not been reported prior to this study. The species in this study all possess the same chromosome number, $n = 36$, and on the basis of $x = 12$, these species are hexaploid. The meiotic behavior in all species, however, is diploid in basic pattern, with no multivalents. The meiotic divisions revealed perfect pairing, with no traces of abnormality.

Experimental Hybridizations

1. General information.—Over 350 experimental hybridizations were attempted between 17 populations of 13 species, these representing a total of 73 different interspecific combinations (fig. 10).

   Bidens species are protandrous and self-compatible. The pollen is available shortly after, or before the buds open. After elongation of the style, the stigmatic surfaces are exposed for 1–3 days, but sometimes the stigmatic surfaces may not become receptive until 1–2 days after anthesis. As the florets do not all open at the same time, it is necessary to repeat the hand-pollination of flowers in the same capitulum several times during succeeding days. Very often it is necessary to remove the unopened floret buds from the capitulum to prevent accidental selfing. In certain species such as *B. wiebkei*, the anthers contain so little pollen that it is difficult to use them as the staminate parent.

   Plants of twenty-three different interspecific hybrids were obtained between sixteen populations of twelve species (fig. 11). Of these, thirteen different interspecific crosses were fertile through the F_2 generation. Some ninety-five F_2 plants were raised from achenes obtained from a single F_1 interspecific hybrid plant (*Bidens hillebrandiana* × *B. ctenophylla*).

2. Cytology of F_1 hybrids.—Cytological study was made of the experimental hybrids by the method used for studying the parental species.

   The experimental hybrids usually have thirty-six pairs, as in the parental species, and all have a "diploid" meiosis. The chromosomes usually show close "diploid" pairing at meiosis, except in two hybrids where univalents were found. A few univalents were observed in the F_1 hybrid between *Bidens coartata* and *B. menziesii*, yet this cross produced a vigorous F_2 generation. In the F_1 hybrid between *B. macrocarpa* and *B. menziesii*, some microsporocytes had up to four univalents.

3. Description and fertility of F_1 hybrids.

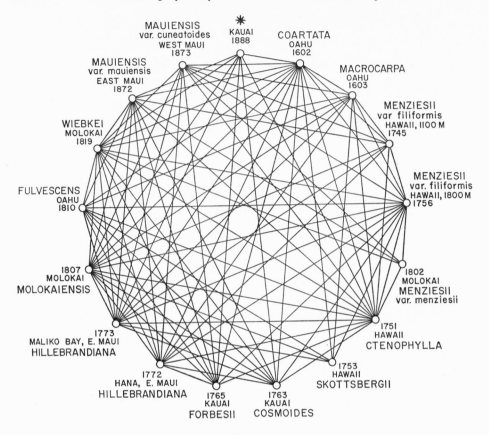

POLYGON SHOWING SPECIES AND LOCALITIES OF
HAWAIIAN <u>BIDENS</u> AND PROGRAM OF ATTEMPTED
EXPERIMENTAL CROSSES.

Fig. 10. Experimental hybridizations attempted between Hawaiian species of *Bidens*.
* Putative hybrid, *Bidens forbesii × B. menziesii.*

Bidens coartata (Gillett 1602) × B. menziesii
var. *filiformis (Gillett 1756)*

(Fig. 2)

The parental species occur on separate islands and in extremely different habitats.
They also differ greatly in habit and leaf form: *Bidens coartata* is less woody with
3- to 5-parted cauline leaves while *B. menziesii* is more woody and with finely dissected cauline leaves.

1602 ♀ × 1756 ♂.—The F_1 hybrids were vigorous and produced about 35 percent mature achenes from self-pollination, and these gave 17.3 percent germination. The eight-month-old F_1 hybrid plants grew to about 0.8 m high. Their habit

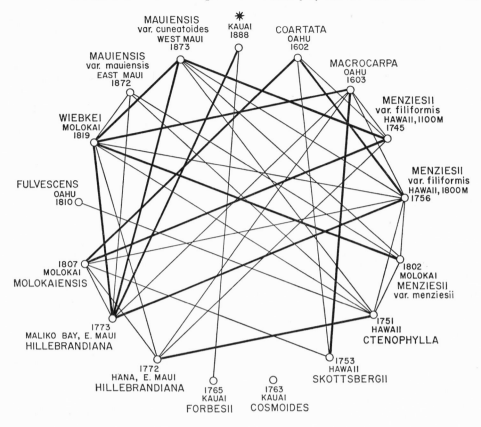

CROSSING POLYGON SHOWING F_1 AND F_2 HYBRIDS

F_1 HYBRIDS ——————

F_2 HYBRIDS ——————

Fig. 11. Successful crosses obtained from experimental cross-pollinations between Hawaiian species of *Bidens*.

* Putative hybrid, *Bidens forbesii* × *B. menziesii*.

was woody, with branches arising at about 35 degrees from the main stem, similar to the habit of *Bidens menziesii*, but with branches more spreading near the ends, similar to those of *B. coartata*. The cauline leaves are 5- to 7-parted, and even bipinnately divided, intermediate between those of the parents (fig. 21). The average width of the segments is also intermediate between those of the parents. They have long acuminate apices and irregularly and deeply dentate margins. The loose corymbose inflorescences are similar to those of *B. coartata*. The capitula are about 2.4 cm in diameter at anthesis. Each capitulum has 5 ray florets and 17–19 disc florets. The style and anther apices are acuminate. The achenes are

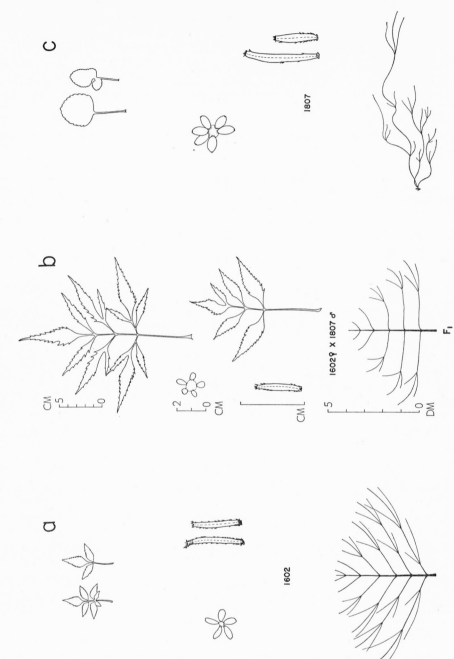

Fig. 12. *a, Bidens coartata; b,* experimental F₁ hybrid, *B. coartata × B. molokaiensis; c, B. molokaiensis.*

black and have narrow lateral marginal wings that are sparsely ciliate. The achenes are 6–9 mm long and 1–1.5 mm broad. The two subapical awns are retrorsely barbed, but of irregular length, from 0.1–1 mm long.

1756 ♀ × *1602* ♂.—The F_1 hybrids from these crosses have the same morphological characters as the reciprocal crosses. They were also vigorous and had about the same fertility as the above F_1 hybrids.

Bidens coartata (*Gillett 1602*)
× *B. molokaiensis* (*Gillett 1807*)
(Fig. 12)

The parental species occur on separate islands on extremely different habitats. Their habit and leaf form are also different: *Bidens coartata* is an erect shrub with 3- to 5-parted cauline leaves while *B. molokaiensis* is a low, diffused, procumbent herb with mainly simple, entire cauline leaves. The F_1 hybrids were vigorous and produced mature achenes from self-pollination. The six-month-old hybrid plants grew erect to a height of about 0.5 m. Their habit was similar to that of *B. coartata* but with the lower branches arising at right angles to the main stem. The cauline leaves are mainly 5- to 7-parted and incompletely bipinnately divided. The segments have long acuminate apices and irregularly and deeply dentate margins. The inflorescences are similar to those of *B. molokaiensis*, each with 1–3 capitula held on long peduncles. The capitula are about 2 cm in diameter at anthesis. Each capitulum consists of 5 ray florets and 16–19 disc florets. The style and anther apices are acute. The achenes are black and are up to 7 mm long and 1 mm broad. They are sparsely ciliate along the lateral margins near the apex and base. The two subapical awns are very short.

Bidens coartata (*Gillett 1602*) × *B. skottsbergii* (*Gillett 1753*)
(Fig. 13)

The parental species occur on two separate islands in entirely different habitats. They also differ in habit and leaf form: *Bidens coartata* is less woody and with 3- to 5-parted cauline leaves while *B. skottsbergii* is more woody and with simple cauline leaves. The F_1 hybrids showed good vigor and produced about 25 percent normal achenes from self-pollination. The eight-month-old hybrid plants grew to about 0.8 m high. They had the woodiness and habit of *B. skottsbergii*, except that the lower branches arose at right angles from the main stem. The cauline leaves are mainly simple, similar to those of *B. skottsbergii*, but with acuminate apices and truncate bases (fig. 14). The margins are intermediate between those of the parents, with inflexed teeth. The inflorescences are loose with few capitula, similar to those of *B. skottsbergii*. The capitula are about 3 cm in diameter at anthesis. Each capitulum consists of 5–6 ray florets and 15–28 disc florets; the ray florets are often partly or completely tubular. The achenes are black, straight or slightly twisted, and broad. They are smooth and are 1–1.5 cm long and 1.5 mm broad. The two apical awns, up to 2 mm long, are strongly barbed and divergent, similar to those of *B. skottsbergii*. The F_2 hybrids were vigorous plants.

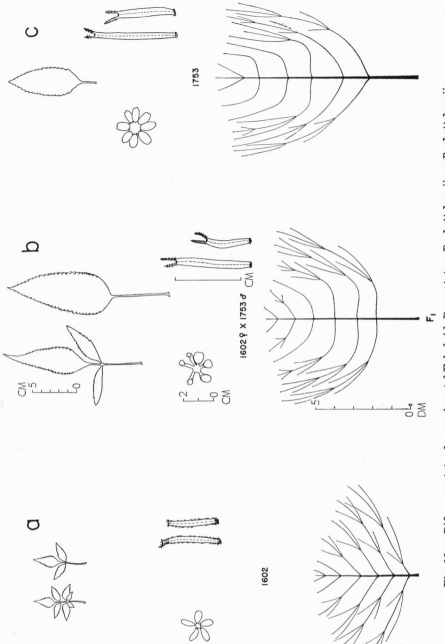

Fig. 13. *a*, *Bidens coartata*; *b*, experimental F₁ hybrid, *B. coartata* × *B. skottsbergii*; *c*, *B. skottsbergii*.

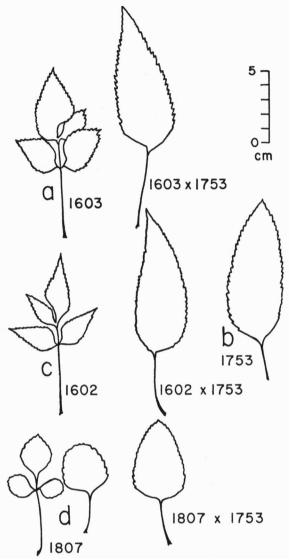

Fig. 14. Inheritance of leaf form in experimental F₁ hybrids of Hawaiian *Bidens. a, B, macrocarpa; b, B. skottsbergii; c, B. coartata; d, B. molokaiensis.*

Bidens hillebrandiana (*Gillett 1772*) × *B. ctenophylla* (*Gillett 1751*)

(Fig. 4)

The parental species occur on separate islands in very different habitats. The parents differ greatly in habit and leaf form. Their F₁ hybrids, however, were vigorous and fertile. From self-pollination, these hybrids set about 95 percent viable achenes and these gave about 99 percent germination. The six-month-old F₁ hybrid plants grew to about 0.6 m high. Their habit was erect with a branching pattern similar to that of *Bidens hillebrandiana*. The cauline leaves are simple, similar to those of *B. ctenophylla* (fig. 15; fig. 16, *a, b*). The leaf apices are acuminate and the bases cuneate; the margins are irregularly and deeply crenate. The capitula are in loose corymbose inflorescences, similar to those of *B. hillebrandiana*. The capitula are about 2.2 cm in diameter at anthesis. Each capitulum consists of 5 ray florets and 8–16 disc florets. The style and anther apices are rather acute, similar to that of the seed parent. The achenes are dark gray and slender, similar to those of *B. hillebrandiana*. They are 9–15 mm long and 1–1.5 mm broad, and are only sparsely ciliate on the surface and near and on the apex. There are usually two, sometimes one or three, apical awns up to 2 mm long. The awns are retrorsely barbed and somewhat divergent.

Ninety-five F_2 plants were raised from F_2 achenes obtained from a single F_1 hybrid plant of *Bidens hillebrandiana (1772)* × *B. ctenophylla (1751)* from self-pollination. These plants were healthy and showed good vigor. They varied in their branching patterns from a virgate main stem devoid of any lateral branches to a deliquescent form. The leaf forms are highly varied, embracing a whole range of expression from simple to finely bipinnately dissected blades (fig. 15). In figure 15, each leaf represents a single plant and was taken from approximately the eighth node above the cotyledons. The leaf forms resemble those of each parental species with a series of intermediate ones. At the lower left-hand corner of figure 15 is a group of finely dissected leaves that resemble the leaf form of *B. menziesii* var. *filiformis*. Furthermore, many of the leaf forms of the intermediate F_2 plants resemble leaves of F_1 hybrids of other combinations.

The above transgressive inheritance clearly indicates that leaf form is regulated by multiple factors. These genes must be few in number but with relatively broad effects since the segregation of both parental types is indicated in the relatively small population of only fifty-six F_2 hybrids.

Bidens hillebrandiana (Gillett 1773) × *B. mauiensis*
var. *cuneatoides (Gillett 1873)*
(Fig. 6)

The parental species were collected from the same island but from different habitats. They differ in habit and leaf form: *Bidens hillebrandiana* is a low, erect plant with dissected leaves while *B. mauiensis* is low, diffuse, and procumbent with mainly simple and thicker cauline leaves. The F_1 hybrids exhibited hybrid vigor and produced normal mature achenes from self-pollination. The seven-month-old hybrid plants grew to 0.4 m high. Their habit was similar to that of *B. hillebrandiana;* erect, with the branches arising at about 70 degrees with the main stem. The cauline leaves are 5-parted with segments shaped more like that of *B. hillebrandiana* but with less deeply cut margins. The inflorescences are simple, as in *B. mauiensis* var. *cuneatoides*, each with 1–3 capitula held by a long peduncle. The capitula are about 2.5 cm in diameter at anthesis and each consists of 5 ray and 16–19 disc florets. The achene structure is intermediate between that of the parents. Achenes are dark gray in color and are up to 1.3 cm long and 2 mm broad. The lateral wings are well covered with upward-pointing hairs. These wings gradually narrow near the apex into two broad and retrorsely barbed awns that are up to 2 mm long.

The F_2 plants were vigorous.

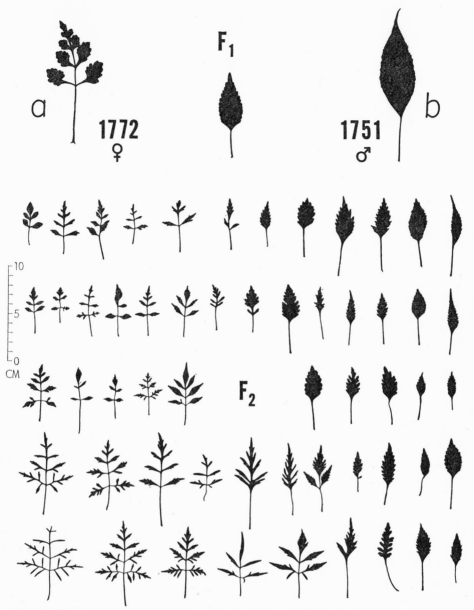

Fig. 15. Segregation of leaf form in an F_2 population of 56 plants of the cross between: *a*, *Bidens hillebrandiana* (Gillett 1772); and, *b*, *B. ctenophylla* (Gillett 1751). Leaves taken from the eighth node above the cotyledons.

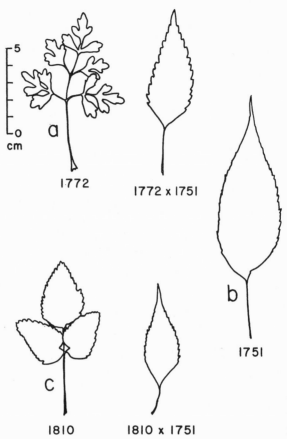

Fig. 16. Inheritance of leaf form in experimental F₁ hybrids of Hawaiian *Bidens. a, B. hille-brandiana; b, B. ctenophylla; c, B. fulvescens.*

Bidens macrocarpa (Gillett 1603) × B. menziesii var. *filiformis (Gillett 1756)*
(Fig. 5)

The parental species occur on separate islands in extremely different habitats: *Bidens macrocarpa* grows in dense rain forest on Oahu while *B. menziesii* var. *filiformis* grows on hot, arid, and exposed upland areas of Hawaii. The species also differ in habit and leaf form: *B. macrocarpa* is less woody with 3- to 5-parted cauline leaves while *B. menziesii* is an erect, robust, woody shrub with finely dissected cauline leaves. The F₁ hybrids exhibited hybrid vigor and produced mature achenes from self-pollination. The six-month-old hybrid plants grew to about 0.5 m high. Their habit was woody, similar to that of *B. menziesii*, but with lower branches arising horizontally at right angles from the main stem, similar to those of *B. macrocarpa*. The cauline leaves are 5- to 7-parted, and even incompletely bipinnately divided, intermediate between those of the parents (fig. 21). The width of the segments is also intermediate between those of the parents. The margins are less closely dentate than those of *B. macrocarpa* and the teeth are reflexed. The loose corymbose inflorescences are similar to those of *B. macrocarpa*. The capitula are about 2.5 cm in diameter at anthesis and each capitulum has 4–5 ray florets and 12–19 disc florets. The ray florets are often tubular. The style apices are acuminate; the another apices are acute. The achenes are dark brown and broad with marginal wings. The achene surface is smooth with a ciliate apex below which a tooth-like outgrowth often arises from the wings. The achenes are 7–9 mm long and about 2 mm broad. The two, sometimes one, retrorsely barbed awns are of irregular lengths, up to 1 mm long, and are curved.

Bidens menziesii var. *filiformis* (*Gillett 1756*) × *B. hillebrandiana* (*Gillett 1773*)
(Fig. 17)

The parental species were collected from different islands and on very different habitats. The species differ in habit but are somewhat similar in their leaf form: *Bidens hillebrandiana* (*Gillett 1773*) is a low herb with dissected cauline leaves while *B. menziesii* var. *filiformis* (*Gillett 1756*) is a tall, erect, woody shrub with very finely dissected leaves.

1773 ♀ × *1756* ♂.—The F_1 hybrids are rather difficult to distinguish on leaf form alone, the leaves being similar to those of *B. hillebrandiana*. They were vigorous and produced about 80 percent mature achenes from self-pollination, and these gave about 50 percent germination, producing vigorous F_2 hybrids. The seven-month-old F_1 hybrid plants grew to 0.7 m high. They were erect and with a branching pattern similar to that of *Bidens menziesii*, with the lower branches arising at about 25 degrees from the main stem, the tips of these level with those of the upper branches, but more spreading. The form and width of the leaf segments is intermediate between those of the parents. The loosely corymbose inflorescences are similar to those of *B. hillebrandiana*. The capitula are about 2.5 cm in diameter at anthesis. Each capitulum consists of 6–7 ray florets and 14–16 disc florets. The style and anther apices are acuminate. The achene structure is similar to that of *B. hillebrandiana* with a more varied number of awns, from 2–6. The awns are up to 2 mm long, strongly and retrorsely barbed [see achenes of *B. hillebrandiana* (*Gillett 1772*) ♀ × *B. menziesii* var. *filiformis* (*Gillett 1745*) ♂, (fig. 23, *c*)]. The achene surface has ridges and these and the lateral margins are covered by upward-pointing hairs. The achenes are dark gray and are 7–9 mm long and about 1.5 mm broad.

1756 ♀ × *1773* ♂.—The F_1 hybrids from these reciprocal crosses do not show any differences in morphological characters except for achene structure (fig. 17, *b*). They were also vigorous and had about the same fertility as the above F_1 hybrids. But the achenes of these hybrids show the expression of matricliny, being more smooth, with two distinctly shorter, unequal but strongly retrorsely barbed awns. Some hybrid achenes are without awns, as in the female parent, *B. menziesii*.

Bidens menziesii var. *filiformis* (*Gillett 1745*) × *B. mauiensis*
var. *cuneatoides* (*Gillett 1873*)
(Fig. 18)

The parental species were collected from different islands, from slightly different habitats. They differ in habit and leaf form: *Bidens menziesii* var. *filiformis* is a

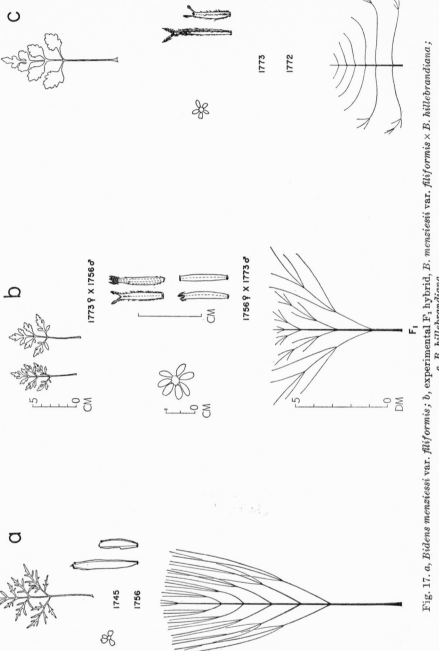

Fig. 17. *a, Bidens menziesii* var. *filiformis*; *b*, experimental F_1 hybrid, *B. menziesii* var. *filiformis* × *B. hillebrandiana*; *c, B. hillebrandiana*.

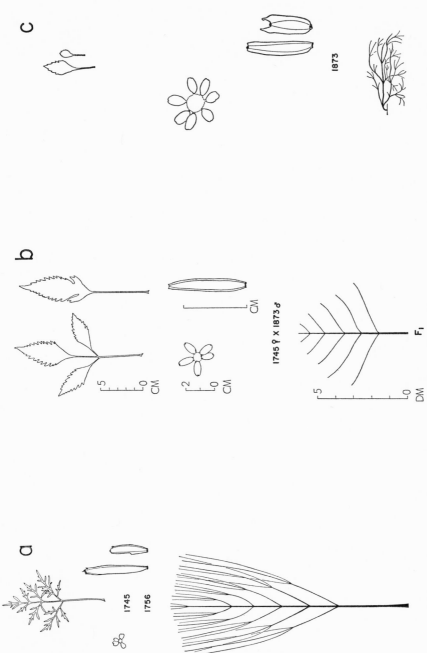

Fig. 18. *a*, *Bidens menziesii* var. *filiformis*; *b*, experimental F₁ hybrid, *B. menziesii* var. *filiformis* × *B. mauiensis* var. *cuneatoides*; *c*, *B. mauiensis* var. *cuneatoides*.

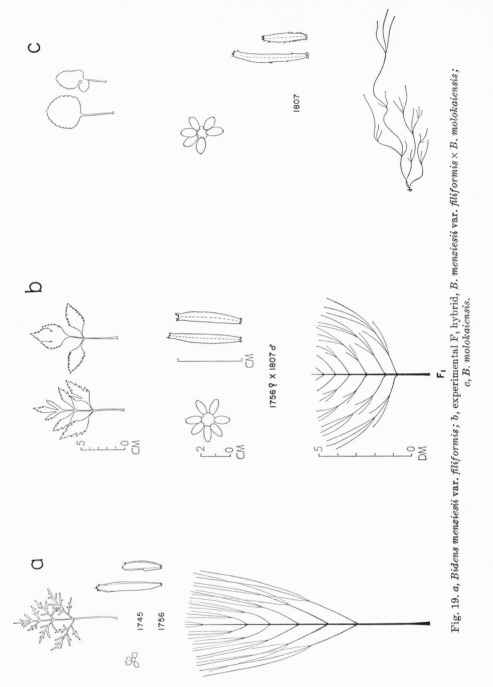

Fig. 19. *a, Bidens menziesii* var. *filiformis*; *b*, experimental F₁ hybrid, *B. menziesii* var. *filiformis* × *B. molokaiensis*; *c, B. molokaiensis*.

tall, woody shrub with finely dissected leaves while *B. mauiensis* var. *cuneatoides* (*Gillett 1873*) is a low, diffuse, procumbent herb with mainly simple and thicker cauline leaves. The F_1 hybrid plants were quite vigorous but produced few achenes per capitulum. These achenes, however, produced vigorous F_2 hybrids. The seven-month-old F_1 hybrid plants grew to 0.5 m high. They were woody and erect with the simple lateral branches arising at about 70 degrees from the main stem. The cauline leaves are mostly 3-parted but with 1–3 simple ones that are deeply cut at the margins and much larger in size. The inflorescences are similar to those of *B. mauiensis* var. *cuneatoides*, with long peduncles. The capitula are about 2.2 cm in diameter at anthesis. Each capitulum consists of 5 ray and 10–14 disc florets. The achene structure is similar to that of *B. mauiensis* var. *cuneatoides* but with narrower lateral wings and greater length than both parents display. They are up to 1.25 cm long and 2.2 mm broad.

Bidens menziesii var. *filiformis* (*Gillett 1745, 1756*)

× *B. molokaiensis* (*Gillett 1807*)

(Fig. 19)

1745 ♀ × *1807* ♂ ; *1756* ♀ × *1807* ♂.—The parental species were collected from different islands, the two races of *Bidens menziesii* var. *filiformis* from different habitats on the island of Hawaii. The habit and leaf form of the parental species are extremely different: *B. menziesii* is a tall, erect, woody shrub with finely dissected cauline leaves while *B. molokaiensis* is a low, diffuse, procumbent herb with mainly simple cauline leaves. The F_1 hybrids from the two different crosses did not show any morphological differences. They were vigorous but both produced very few mature achenes from self-pollination. Vigorous F_2 hybrids (*1745* × *1807*), however, were produced. The seven-month-old F_1 hybrid plants grew to a height of 0.65 m. Their habit was similar to that of *B. menziesii* var. *filiformis* but with branches arising at greater angles (about 70 degrees) from the main stem. The cauline leaves are intermediate between those of the parents. They are mainly 3- to 5-parted, and even incompletely bipinnately divided with the width of segments intermediate. The margins are irregularly and deeply dentate. The inflorescences are similar to those of *B. molokaiensis*, each with 1–3 capitula held by a long peduncle or peduncles. The capitula are about 3 cm in diameter at anthesis. Each capitulum consists of 5–7 ray florets and 13–15 disc florets. The style apices are acuminate and the anther apices obtuse. The achenes are dark brown with a smooth surface except at the apices. They are 7–12 mm long and 1–1.5 mm broad. There are two very short, barbless awns, and sometimes these are missing.

Bidens molokaiensis (*Gillett 1807*) × *B. hillebrandiana* (*Gillett 1773*)

(Fig. 20)

The parental species were collected from different islands, but on somewhat similar habitats. Their habit and leaf form are very different: *Bidens molokaiensis* is a low, diffuse, procumbent herb with usually simple cauline leaves while *B. hillebrandiana* is an erect herb with strongly lobed and dissected cauline leaves.

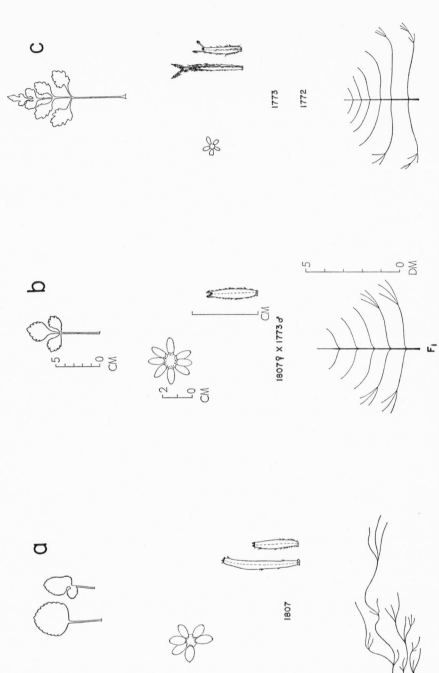

Fig. 20. *a, Bidens molokaiensis; b,* experimental F₁ hybrid, *B. molokaiensis × B. hillebrandiana; c, B. hillebrandiana.*

1807 ♀ × *1773* ♂.—The F_1 hybrids were vigorous and produced a few mature achenes from self-pollination. The six-month-old hybrid plants grew to 0.5 m high. They were erect and with branches arising at almost right angles from the main stem, similar to those of *Bidens hillebrandiana*, except that the lower branches were not reclining. This habit contrasts with that of both parents. The cauline leaves are 3-parted with the width of segments and margins intermediate between those of the parents. The inflorescences are similar to those of *B. molokaiensis* each with 1–3 capitula held on a long peduncle or peduncles. The capitula are about 3 cm in diameter at anthesis. Each capitulum consists of 6–10 ray florets and 20–40 disc florets. The style apices are acute but the anther apices are obtuse. The achenes are black and are 8–10 mm. long and about 1.5 mm broad. The achene surface is smooth but the lateral margins and apex are ciliate. The two apical, short awns, 1–1.5 mm long, are retrorsely barbed.

1773 ♀ × *1807* ♂.—The F_1 hybrids from these reciprocal crosses did not differ morphologically from the above F_1 hybrids. They were vigorous and had about the same fertility as the above F_1 hybrids.

Bidens molokaiensis (*Gillett 1807*) × *B. skottsbergii* (*Gillett 1753*)
(Fig. 7)

The parental species occur on separate islands in different habitats. They are extremely different in habit: *Bidens molokaiensis* is a low, diffuse, procumbent herb while *B. skottsbergii* is an erect, woody shrub. They both have simple cauline leaves but of different form. The F_1 hybrids showed vigor but produced only about 10 percent mature achenes from self-pollination. The seven-month-old F_1 hybrid plants grew to 0.7 m high. Their habit was erect and woody, similar to that of *B. skottsbergii*, except that the lower branches diverged at right angles from the main stem and the tops of the branches were not in one plane. The cauline leaves are simple with their form intermediate between that of the parents (figs. 7, 14). The inflorescences are similar to those of *B. molokaiensis*, each with 1–3 long-pedunculate capitula. The capitula are about 3.2 cm in diameter at anthesis. Each capitulum consists of 7–8 ray florets and 20–40 disc florets. The ray florets often are tubular. The achenes are similar to those of *B. skottsbergii*. They are sparsely ciliate along the lower lateral margins and are 0.9–1.2 cm long and 1.5 mm broad. The two long apical awns, up to 2 mm long, are retrorsely barbed and divergent, showing epistasis for the awned genes.

Bidens wiebkei (*Gillett 1819*) × *B. mauiensis* var. *cuneatoides* (*Gillett 1873*)
(Fig. 9)

The parental species were collected from two different islands on somewhat similar habitats. Their habit and leaf form are very different: *Bidens wiebkei* is a tall woody shrub with 3- to 5-parted cauline leaves while *B. mauiensis* is a low, diffused, procumbent herb with mainly simple and thicker cauline leaves. The F_1 hybrid plants were rather vigorous and produced many mature achenes from self-pollination. These germinated readily and produced a vigorous F_2 generation. The seven-month-old F_1 hybrid plants grew to 0.5 m high. Their habit was woody and erect, similar to that of *B. wiebkei*, with their simple branches arising at about

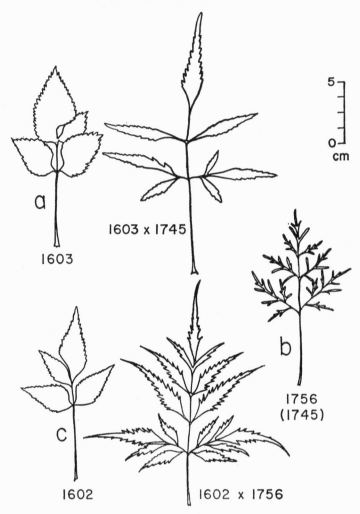

Fig. 21. Inheritance of leaf form in experimental F_1 hybrids of Hawaiian *Bidens*. *a, B. macrocarpa; b, B. menziesii* var. *filiformis; c, B. coartata.*

90 degrees from the main stem. Their cauline leaves are similar to those of *B.
wiebkei.* The inflorescences are similar to those of *B. mauiensis* var. *cuneatoides*
each consisting of 1–3 long-pedunculate capitula. The capitula are about 2.8 cm
in diameter at anthesis. Each capitulum consists of 5–7 ray florets and 16–18
disc florets. The achene structure is intermediate between that of the parents (fig.
23, *d*). The achenes are dark brown in color and are 7–8 mm long and 2 mm broad.
The marginal wings and apices are covered with upward pointing hairs and two
apical awns extend from the wings. The two awns, up to 1.5 mm long, are retrorsely
barbed.

Genotypic Characters

In comparing the morphological characters of F_1 hybrids with those of their
parents, the following inheritance patterns are noticeable:

1. Woody and erect habit.—This habit is to be noted in the F_1 hybrids between woody, erect species and procumbent, herbaceous species in the following: *Bidens coartata* × *B. molokaiensis* (fig. 12); *B. hillebrandiana* × *B. mauiensis* var. *cuneatoides* (fig. 6); *B. hillebrandiana* × *B. molokaiensis* (fig. 20); *B. menziesii* var. *filiformis* × *B. mauiensis* var. *cuneatoides* (fig. 18); *B. menziesii* var. *filiformis* × *B. molokaiensis* (fig. 19); *B. molokaiensis* × *B. skottsbergii* (fig. 7); and *B. wiebkei* × *B. mauiensis* var. *cuneatoides* (fig. 9). The woody and erect habit thus is seen to be regulated by epistatic genes.

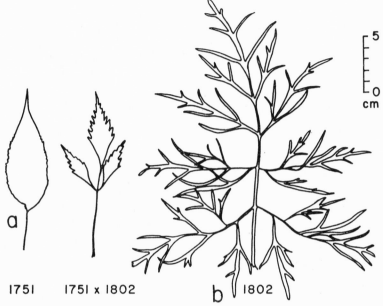

Fig. 22. Inheritance of leaf form in Hawaiian *Bidens*. *a*, *B. ctenophylla*; *b*, *B. menziesii* var. *menziesii*.

2. Simple and bipinnately dissected leaf forms:
    a. Simple.—This form is more strongly expressed in the F_1 hybrids than the bipinnately dissected and pinnately dissected leaf form as in the following: *B. hillebrandiana* × *B. ctenophylla* (fig. 4); *B. molokaiensis* × *B. skottsbergii* (fig. 7); *Bidens coartata* × *B. skottsbergii* (fig. 13); *B. macrocarpa* × *B. skottsbergii* (fig. 14, *a*, *b*); *B. fulvescens* × *B. ctenophylla* (fig. 16, *b*, *c*); and *B. ctenophylla* × *B. menziesii* var. *menziesii* (fig. 22). Thus the simple expression appears to be regulated by genes that are epistatic over the bipinnately dissected and pinnately dissected expressions.
3. Simple inflorescence.—The inheritance of inflorescence characters is shown in table 2. When a species with a loosely corymbose inflorescence is crossed with a species with compactly corymbose inflorescence, the loosely corymbose condition is expressed in their F_1 hybrids. When a species with a solitary inflorescence is crossed with a species with a loosely corymbose or a compactly corymbose inflorescence, the solitary condition is usually expressed in the F_1. Thus the loosely corymbose condition appears to be regulated by genes that are

epistatic over the compactly corymbose and the genes regulating the solitary condition appear to be relatively epistatic over loosely and compactly corymbose conditions.

4. Awned.—The inheritance of achene awns is noted in table 2 and figure 23. When a species with awned achene is crossed with a species with vestigially awned achene or with awnless achene, the awned condition is found in their F_1 hybrids (fig. 23; figs. 4, 5, 6, 7, 9, 13, 17, 20).

TABLE 2

INHERITANCE OF INFLORESCENCE AND ACHENE AWNS
IN PARENTAL SPECIES OF BIDENS AND THEIR F_1 HYBRIDS

Female Male	Hybrids	Parents	
		Female	Male
coartata x *menziesii* var. *filiformis*	LA	LA	Ca
menziesii var. *filiformis* x *coartata*	LA	Ca	LA
coartata x *molokaiensis*	SA	LA	SA
coartata x *skottsbergii*	LA̲	LA	LA
hillebrandiana x *ctenophylla*	LA̲	LA	CA̲
hillebrandiana x *mauiensis* var. *cuneatoides*	SA̲	LA	Sa
macrocarpa x *menziesii* var. *filiformis*	LA̲	LA	Ca
menziesii var. *filiformis* x *hillebrandiana*	LA̲	Ca	LA̲
hillebrandiana x *menziesii* var. *filiformis*	LA̲	LA̲	Ca
menziesii var. *filiformis* x *mauiensis* var. *cuneatoides*	Sa	Ca	Sa
menziesii var. *filiformis* x *molokaiensis*	SA	Ca	SA
molokaiensis x *hillebrandiana*	SA̲	SA	LA̲
hillebrandiana x *molokaiensis*	SA̲	LA̲	SA
molokaiensis x *skottsbergii*	SA̲	SA	LA̲
wiebkei x *mauiensis* var. *cuneatoides*	SA̲	CA̲	Sa

Inflorescence Expression	Expression of Awns in Achenes
S = Solitary	A = Awned
L = Loosely corymbose	A̲ = Vestigially awned
C = Compactly corymbose	a = awnless

BIOCHEMICAL COMPARISON OF THE PARENTAL SPECIES AND THEIR F_1 HYBRIDS

It has been stated by Alston (1967) that phenolic secondary compounds actually have an evolutionary history, are regulated genetically, and so can be used as criteria for evolutionary affinity and taxonomic relationship. These compounds play an important role in revealing biochemical relationships mainly because they apparently do not take part in the basic metabolism of the cell and are in general of less importance to the organisms and thus chemical variation may follow. Phenolic secondary compounds are numerous, chemically diverse, and are stable in ordinary extraction methods. Because of these advantages, they were chosen for study.

The numbers of chromatographic spots of phenolic compounds in this material ranges from twenty-two to thirty-eight per species. A total of seventy-one different spots appeared on the chromatograms of the species studied (table 3).

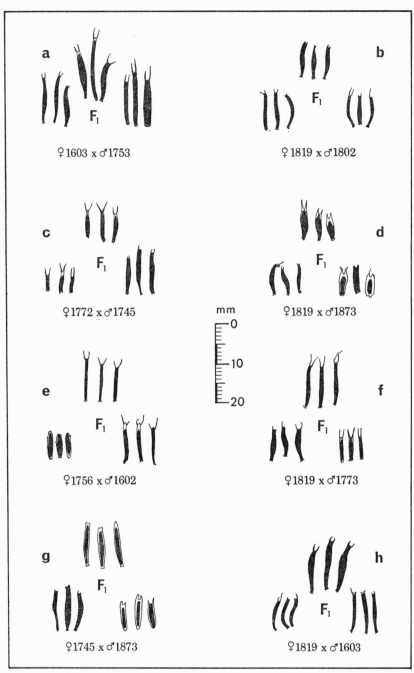

Fig. 23. Inheritance of achene structure in experimental interspecific crosses of Hawaiian *Bidens. a, B. macrocarpa* (1603) × *B. skottsbergii* (1753); *b, B. wiebkei* (1819) × *B. menziesii* var. *menziesii* (1802); *c, B. hillebrandiana* (1772) × *B. menziessi* var. *filiformis* (1745); *d, B. wiebkei* (1819) × *B. mauiensis* var. *cuneatoides* (1873); *e, B. menziesii* var. *filiformis* (1756) × *B. coartata* (1602); *f, B. wiebkei* (1819) × *B. hillebrandiana* (1773); *g, B. menziesii* var. *filiformis* (1745) × *B. mauiensis* var. *cuneatoides* (1873); *h, B. wiebkei* (1819) × *B. macrocarpa* (1603).

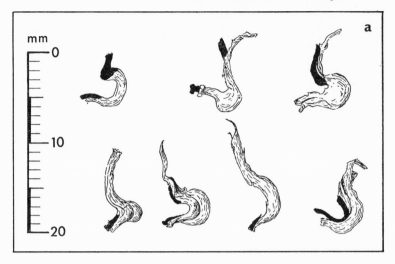

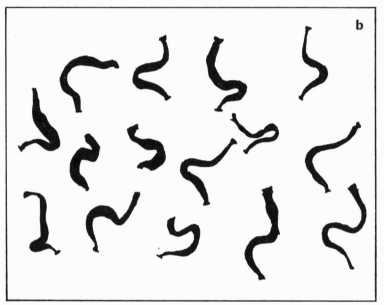

Fig. 24. Mature achenes of *Bidens cosmoides* (A. Gray) Sherff. *a*, achenes with clasping abaxial bract; *b*, achenes with clasping abaxial bract removed.

Six spots were detected in all of the species and their experimental hybrids (table 3; spots 1, 2, 8, 14, 21, 38). Of the seventy-one spots, eleven appeared in each instance only in a single species, and eight in not more than two species (tables 3, 4). *Bidens coartata* has three unique spots (tables 3, 4: nos. 42, 44, 67); *B. cosmoides* and *B. menziesii* each have two unique ones (tables 3, 4: nos. 27, 37; 22, 53, respectively); *B. forbesii, B. macrocarpa, B. skottsbergii* and *B. fulvescens* each have one spot of its own (tables 3, 4: nos. 12, 3, 63, 66, respectively). *Bidens menziesii* shares two of its spots with *B. coartata* (tables 3, 4: nos. 45, 48), and four spots with *B. forbesii, B. skottsbergii, B. hillebrandiana* and *B. mauiensis* var. *mauiensis* (1872) (tables 3, 4: nos. 23, 10, 51, 41, respectively). *Bidens fulvescens* shares two of its spots with *B. hillebrandiana* and *B. mauiensis* var. *cuneatoides* (1873), respectively (tables 3, 4: nos. 4, 56, respectively).

PAIRED AFFINITY INDICES

In recent years, much emphasis has been placed on the possible use of the chromatographic pattern as an indicator of taxonomic affinity. It was emphasized that such chromatographic data could be utilized without actual knowledge of the

TABLE 4

CHROMATOGRAPHIC SPOTS WHICH ARE MORE SPECIFIC
IN THEIR DISTRIBUTION AMONG THE SPECIES STUDIED

	Spot number	
	In a single species	Shared by 2 species
coartata	42, 44, 67	45, 48
cosmoides	27, 37	
menziesii var. *filiformis*	22, 53	45, 48, 23 ,10, 51, 41
forbesii	12	23
macrocarpa	3	
skottsbergii	63	10
fulvescens	66	4, 56
hillebrandiana		51 4
mauiensis var. *mauiensis 1872*		41
mauiensis var. *cuneatoides 1873*		56

identities of the chemical compounds (Turner and Alston, 1959; Brehm, 1963; Horne, 1965). The concept of paired affinity indices for the representation of chromatographic data was first developed and quite successfully applied to the analysis of *Bahia* by Ellison et al. (1962). This method has also been successfully employed in the study of *Lotus* species by Harney and Grant (1963). It is important to have a large number of spots with a large extent of interspecific variability to work on in order to be able to use this method. The more chemical components that are available for comparison, the more reliable will be the systematic conclusions. Hawaiian species of *Bidens* appear to fit these requirements very well. Each species has at least twenty-two chemical components and there is great chemical variation in the total number of seventy-one spots. This means of comparison can serve as a supplement to other information.

The spots on the chromatograms were observed and recorded in ultraviolet light. Consistent number designations were assigned to spots that appeared to be identical according to color and position in two or more taxa, (table 3; figs. 25, 26, 27). The chromatograms of paired taxa were compared and the numbers of spots shared by each pair and the total number of spots occurring in both taxa of each pair were determined. The ratio of the former to the latter for each pair was expressed as a percentage called the paired affinity index (PA). The PA indices of a given species with all other species were expressed along the radii (each of which repre-

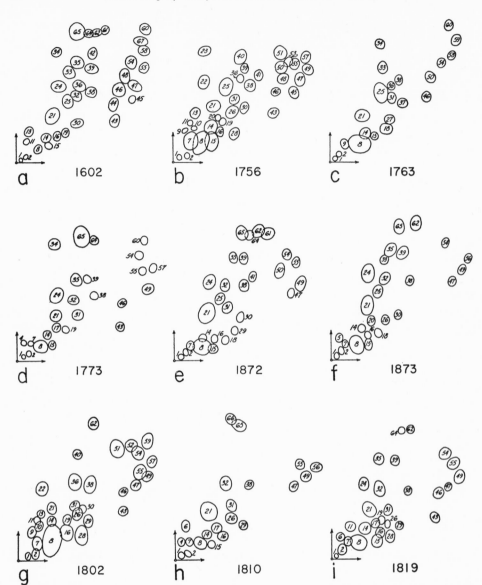

Fig. 25. Chromatograms of polyphenol compounds of Hawaiian species of *Bidens*. Vertical axis represents the first run. Spot numbers as listed in table 3, *a*, *B. coartata*; *b*, *B. menziesii* var. *filiformis*; *c*, *B. cosmoides*; *d*, *B. hillebrandiana*; *e*, *B. mauiensis* var. *mauiensis*; *f*, *B. mauiensis* var. *cuneatoides*; *g*, *B. menziesii* var. *menziesii*; *h*, *B. fulvescens*; *i*, *B. wiebkei*.

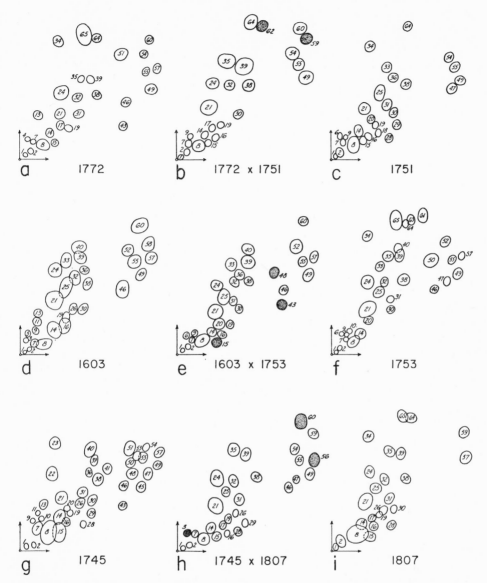

Fig. 26. Chromatograms of polyphenol compounds of Hawaiian species of *Bidens* and experimental F₁ hybrids. Vertical axis represents first run. Spot numbers listed in table 3. Shaded spots in hybrids do not occur in either parent. *a, B. hillebrandiana; b, B. hillebrandiana × B. ctenophylla; c, B. ctenophylla; d, B. macrocarpa; e, B. macrocarpa × B. skottsbergii; f, B. skottsbergii; g, B. menziesii var. filiformis; h, B. menziesii var. filiformis × B. molokaiensis; i, B. molokaiensis.*

sents one taxon) of a circle starting at 0 percent from the center to 100 percent at the periphery. This enables the degree of chemical similarities and differences to be noted at a glance (figs. 28, 29).

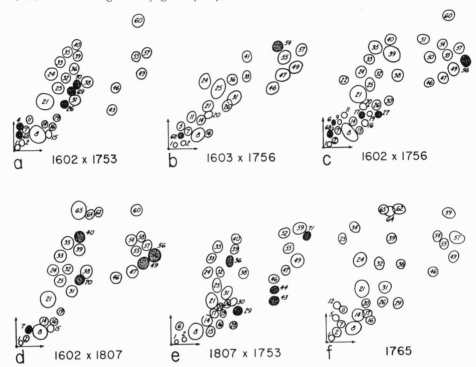

Fig. 27. Chromatograms of polyphenol compounds of Hawaiian species of *Bidens* and experimental F₁ hybrids. Vertical axis represents first run. Spot numbers listed in table 3. Shaded spots in hybrids do not occur in either parent. *a, B. coartata* × *B. skottsbergii; b, B. macrocarpa* × *B. menziesii* var. *filiformis; c, B. coartata* × *B. menziesii* var. *filiformis; d, B. coartata* × *B. molokaiensis; e, B. molokaiensis* × *B. skottsbergii; f, B. forbesii.*

GROUP AFFINITY INDICES

Each polygonal graph was expressed as a numerical index called a group affinity (GA) index that is the summation of all the PA indices of any species with all other species (table 5). This provides a quantitative expression of the systematic relationships among the species. The maximum possible GA value for each of these 15 taxa is 1,500 and the minimum possible GA value is 100 which indicate a 100 percent and 0 percent affinity, respectively, of any taxon with itself and the other fourteen taxa.

According to their PA indices with one another, the following species appear to be more closely related biochemically: *Bidens mauiensis* var. *mauiensis* (1872), *B. wiebkei, B. hillebrandiana, B. molokaiensis,* and *B. forbesii.* They have PA indices of more than 50 percent with one another (with the exception of the PA index between *B. mauiensis* var. *mauiensis* and *B. molokaiensis* which is 49 percent). The polygonal graphs of *B. ctenophylla, B. fulvescens, B. mauiensis* var. *cuneatoides, B. macrocarpa,* and *B. skottsbergii* show their intermediate positions;

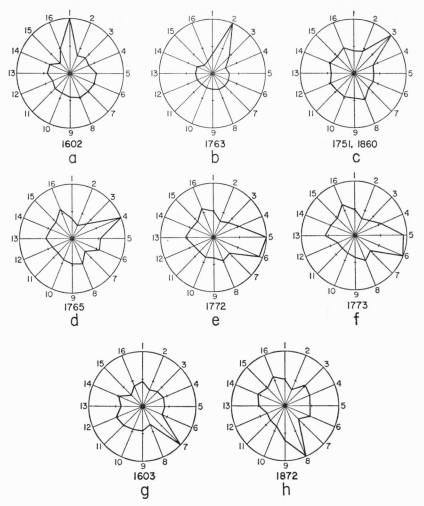

Fig. 28. Paired affinity indices of Hawaiian species of *Bidens* based on chromatograms in figs. 25, 26, 27. Inner hash marks denote an affinity of 50 percent, outer circle denotes 100 percent affinity.

1. *B. coartata* (1602)
2. *B. cosmoides* (1763)
3. *B. ctenophylla* (1751, 1860)
4. *B. forbesii* (1765)
5. *B. hillebrandiana* (1772)
6. *B. hillebrandiana* (1773)
7. *B. macrocarpa* (1603)
8. *B. mauiensis* var. *mauiensis* (1872)

9. *B. mauiensis* var. *cuneatoides* (1873)
10. *B. menziesii* var. *filiformis* (1745)
11. *B. menziesii* var. *filiformis* (1756)
12. *B. menziesii* var. *menziesii* (1802)
13. *B. molokaiensis* (1807)
14. *B. skottsbergii* (1753)
15. *B. fulvescens* (1810)
16. *B. wiebkei* (1819)

a, *B. coartata*; *b*, *B. cosmoides*; *c*, *B. ctenophylla*; *d*, *B. forbesii*; *e*, *B. hillebrandiana* (Hana, Maui); *f*, *B. hillebrandiana* (Maliko Bay, Maui); *g*, *B. macrocarpa*; *h*, *B. mauiensis*. var. *mauiensis*.

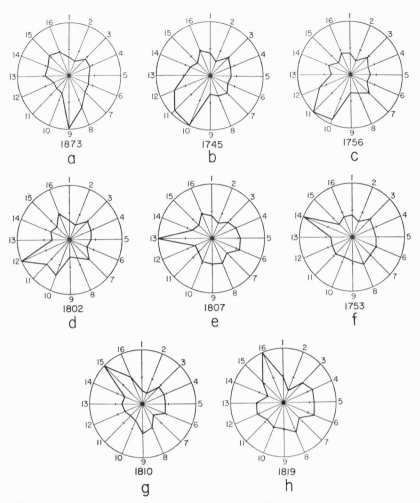

Fig. 29. Paired affinity indices of Hawaiian species of *Bidens,* based on chromatograms in figs. 25, 26, 27. Inner hash marks denote affinity of 50 percent. Outer circle denotes affinity of 100 percent.

1. *B. coartata* (1602)
2. *B. cosmoides* (1763)
3. *B. ctenophylla* (1751, 1860)
4. *B. forbesii* (1765)
5. *B. hillebrandiana* (1772)
6. *B. hillebrandiana* (1773)
7. *B. macrocarpa* (1603)
8. *B. mauiensis* var. *mauiensis* (1872)

9. *B. mauiensis* var. *cuneatoides* (1873)
10. *B. menziesii* var. *filiformis* (1745)
11. *B. menziesii* var. *filiformis* (1756)
12. *B. menziesii* var. *menziesii* (1802)
13. *B. molokaiensis* (1807)
14. *B. skottsbergii* (1753)
15. *B. fulvescens* (1810)
16. *B. wiebkei* (1819)

a, B. mauiensis var. *cuneatoides; b, B. menziesii* var. *filiformis* (Puu Waawaa Ranch, Hawaii); *c, B. menziesii* var. *filiformis* (Saddle Road, Hawaii); *d, B. menziesii* var. *menziesii; e, B. molokaiensis; f, B. skottsbergii; g, B. fulvescens; h, B. wiebkei.*

they have PA indices with all others of between 50 percent and 30 percent with one or two exceptions (figs. 28, 29). *Bidens menziesii* (1745, 1756, 1802) *B. coartata,* and *B. cosmoides* have their PA indices with all others at less than 50 percent. *B. cosmoides* is distinctive in having its PA indices at around 30 percent except one (44 percent with *B. ctenophylla*) and this species has the lowest group affinity index.

All of the species studied have GA values above 684 with the exception of *Bidens cosmoides* (table 5). *Bidens wiebkei* has the highest GA value, 845, and this ap-

TABLE 5

GROUP AFFINITY VALUES

Species	GA
B. wiebkei	845
B. hillebrandiana (average of *1772* and *1773*)	815
B. mauiensis var. *mauiensis* (*1872*)	802
B. menziesii var. *filiformis* (*1745*)	786
B. forbesii	770
B. ctenophylla	766
B. molokaiensis	756
B. menziesii var. *filiformis* (*1756*)	755
B. mauiensis var. *cuneatoides* (*1873*)	753
B. skottsbergii	749
B. menziesii var. *menziesii* (*1802*)	747
B. coartata	725
B. macrocarpa	725
B. fulvescens	684
B. cosmoides	560

pears to agree quite closely with its ease of forming experimental hybrids with most other species. This is also the case with *B. hillebrandiana. Bidens cosmoides* has a GA value (560), which is much lower than the rest, and this coincides with its peculiar morphology and the impossibility of obtaining any experimental hybrids with the other species.

Thin-layer chromatography was done on eight of the experimental interspecific F_1 hybrids and the results are summarized in tables 3 and 4.

For the interspecific hybrids, the total number of spots which appeared in each chromatogram ranged from twenty-four to thirty-seven. A total of fifty-seven different spots were detected on the hybrid chromatograms, and ten of these were common to all the hybrids (table 3).

From tables 3 and 4, it is apparent that many of the compounds found in the parental species are absent in their interspecific hybrids, and that some new compounds are produced in the hybrids. This has been reported to be the case in the hybrids of *Baptisia* (Alston and Simmons, 1962; Alston and Turner, 1964) and *Lotus* (Harney and Grant, 1963) and in other natural hybrids (Alston et al., 1965). The production of new compounds in experimental hybrids has been claimed by Alston (1967), based on the evidence from the analysis and identification of compounds (Alston et al., 1965; Mabry et al., 1965).

The fact that there are compounds in the hybrids which are absent in either parent, but present in other species (table 6) may be an indication that the production of these compounds is governed by multiple factors such as those governing the leaf form. These genes must be few in number but shared by and shuffled among these species, resulting in the reappearance of certain compounds in interspecific hybrids.

TABLE 6

POLYPHENOL COMPOUNDS IN INTERSPECIFIC HYBRIDS
THAT DO NOT OCCUR IN EITHER PARENT

Hybrids	Spot number	
	(Hybrid compounds not found in parents)	
	Found in other species	Not in other species
B. coartata (1602) x *B. menziesii* var. *filiformis* (1756)...............	6, 17, 27, 56, 68	
B. coartata (1602) x *B. molokaiensis* (1807)..	7, 40, 49, 56	70
B. coartata (1602) x *B. skottsbergii* (1753)...	4, 26, 68	69, 70
B. hillebrandiana (1772) x *B. ctenophylla* (1751)....................	59, 62	
B. macrocarpa (1603) x *B. menziesii* var. *filiformis* (1756).................	54, 68	
B. macrocarpa (1603) x *skottsbergii* (1753)...	15, 43, 48	
B. menziesii var. *filiformis* (1745) x *B. molokaiensis* (1807).................	3 56, 60	
B. molokaiensis (1807) x *B. skottsbergii* (1753).....................	29, 36, 43, 44	71

In general, the biochemical information derived from these studies correlates well with morphological and genetical criteria. The biochemical information thus makes possible a more complete picture of attributes and relationships for the genus *Bidens* in the Hawaiian Islands. At the same time, it must be emphasized that, taken by itself, the biochemical data can scarcely provide an adequate picture of relationships and lineages. If one were faced with the problem of identifying the parentage of a putative wild hybrid through biochemical studies, it is most unlikely that any accurate and satisfactory designation of parentage could be obtained. The presence of entirely new compounds in experimental hybrids, the absence (hypostasis) of certain parental compounds, and the presence (epistasis) of other parental compounds suggest that the genetical regulation of these compounds is complex and variable, as is the case with an assemblage of different morphological characters.

NATURAL HYBRIDS

The acceptance of the enumerated hybrid combinations in the following paragraphs is supported by various lines of evidence, including the sympatric occurrence of the parental species and putative natural hybrids, the comparison of putative hybrid specimens with vouchers of experimental hybrids, and the conspicuous variability of the population of natural hybrids.

In some instances the parents and putative hybrids are allopatric, the notable examples being hybrid populations on Kauai and Oahu in which one of the putative parents is *Bidens menziesii,* currently restricted to Hawaii, Maui, and Molokai. Populations of *B. menziesii* clearly are heterozygous for the presence of awns on the achenes, some plants having distinctly awned achenes and others having awnless achenes. Thus the means for the dispersal of this species throughout the Hawaiian Islands clearly is present. The strong relationship of this species to dry, upland habitats suggests that it probably occurred on Oahu and Kauai during the earlier evolutionary history of these islands.

The first five of the following sixteen enumerated putative hybrids were noted by Sherff and are based on collections of intermediate populations that are sympatric with both parental species. Cited collections are at the B. P. Bishop Museum.

1. *Bidens fecunda* × *B. torta* (Sherff, 1937)

OAHU: Makaleha Valley, *Skottsberg 387, Degener 20830;* Makaleha Ridge, *Meebold s.n.;* Keawaula Valley, *Degener* et al. *4101.*

2. *Bidens amplectens* × *B. torta* (Sherff, 1937)

OAHU: Mt. Kaala, *St. John and Fosberg 12178;* Kukuiala Valley, *E. H. Bryan 827;* Manini Gulch, *E. H. Bryan 793;* Puu Hapapa, *Meebold s.n.;* Puu Kaua, *Degener* et al. *10299;* Mokuleia, *Degener 20631;* Makaleha Valley, *Degener* et al. *10047;* Makua Valley, *Degener* et al. *7474;* Kaiwikoele Stream, *Degener 30496;* Paumalu, *Degener 24502.*

3. *Bidens amplectens* × *B. waianensis* (Sherff, 1937)

In our work it is evident that *Bidens waianensis* consists of a broad spectrum of intermediates between *B. fulvescens* and *B. menziesii.* The hybrid formula for the material designated by Sherff as *B. amplectens* × *B. waianensis,* therefore, should be designated as *B. amplectens* × (*B. fulvescens* × *B. menziesii*). The parents and putative hybrids are sympatric.

OAHU: Keawaula Valley, *Degener* et al. *4120;* Kanehoa Valley, *Pearsall 225;* Ridge between Mt. Kaala and Mt. Kalena, *Degener* et al. *4136.*

4. *Bidens mauiensis* var. *cuneatoides* × *B. menziesii* (Sherff, 1941)

This putative hybrid is substantiated by comparisons between herbarium material and voucher specimens of our experimental F_1 and F_2 hybrids between these two taxa (*Gillett 1873* × *1756; 1745* × *1873; 1873* × *1802*). While the comparisons are not exact, they fall well within the expected variability from such a cross. The closer comparison is between the wild material and certain experimental F_2 hybrids. The natural hybrids have a notable variability and are well represented in the herbarium collections. Sherff (1941) cites a mass collection of 200 plants obtained by Degener in the cited locality of Ukumehame Gulch, West Maui, in 1939. These hybrids were reported to be sympatric with both parents at the locality.

WEST MAUI: McGregor, *Degener* et al. *2678, 2679, 22037;* Olowalu, *Nitta 11205-a;* Ukumehame Gulch, *Nitta 11203, 11204.*

LANAI: Maunalei, *Munro 240, 241, 450, 451, 535;* Awaluana, *Degener 28769* (Isotype of *Bidens awaluana* Deg. and Deg. and Sherff, Sherff, 1964).

5. *Bidens torta* × *B. waianensis* (Sherff, 1951)

The formula for this putative hybrid should be *B. torta* × (*B. fulvescens* × *B. menziesii*). The parents and hybrids are sympatric.

OAHU: Puu Kumakalii, *Degener* et al. *10918;* DuPont Trail, *Degener* et al. *20706;* Makaha Gulch, *St. John 25510;* between Puu Kumakalii and Puu Kalena, *Topping 3408-a;* ridge between Mt. Kaala and Mt. Kalena, *Degener* et al. *4135.*

6. *Bidens fulvescens* × *B. menziesii*

This putative hybrid clearly reflects the foliage of *Bidens menziesii* and the achene form of *B. fulvescens*. Distribution of this hybrid lineage centers on Kolekole Pass, Waianae Mts., Oahu. The region is in one of the most arid sections of Oahu and there is a strong resemblance between habitats of the hybrids and those of *B. menziesii* populations on Molokai, Maui, and Hawaii. The hybrid is sympatric with the *B. fulvescens* parent.

OAHU: Kolekhole Pass, *Forbes 2023-0* (Isotype of *B. waianensis* Sherff), *Selling 3943, Stokes s.n. Swezey s.n., Degener* et al. *2294, 2321, 4115, 4130, 4131;* Makaha Valley, *Skottsberg 1135.*

7. *Bidens hillebrandiana* × *B. mauiensis* var. *cuneatoides*

Herbarium specimens of this putative hybrid compare very well with vouchers of experimental F₁ hybrids (*Gillett 1773 × 1873*). While the parents are not sympatric with the putative hybrids, the habit, foliage, and achene form of the hybrids clearly indicate a gene combination involving these two parental lines.

WEST MAUI: Ili O Kukuipuka, *Degener* et al. *19396.*

8. *Bidens hillebrandiana* × *B. wiebkei*

Experimental F₁ and F₂ hybrids (*Gillett 1819 × 1773*) compare very well with several putative hybrid specimens in the herbarium. The hybrids are restricted to Oahu, while the parental species occur on Maui and Molokai. The awned achenes, however, provide the parental species and hybrids with an excellent means of dispersal. The putative hybrids occur on dry slopes on eastern Oahu, the habitats bearing a strong resemblance to the current habitats of both parental species.

OAHU: Waialae Iki Ridge, *Forbes 2435-0* (Type of *Bidens graciloides* Sherff); Kaimuki Ridge, *Forbes 1862-0;* east rim of Manoa Valley, *Degener* et al. *3411;* Ridge between Niu & Wailupe, *Forbes 2474-0, 2477-0;* Mt. Tantalus, *Degener 2095;* Niu Ridge, *Topping 3304, Degener* et al. *4040, 4041;* east ridge of Kuliouou Valley, *Degener* at al. *4045;* middle Waialae Ridge, *Degener* et al. *4091, 4092* (Type of *Bidens sandwicensis* var. *imminuta* Deg. and Sherff); ridge between Palolo Valley and Waialae Nui Valley, *Fosberg 9710;* Manoa Palolo Ridge, *Egler 37-181;* ridge east of Wailupe, *Egler 37-190;* Hawaiiloa Ridge, *Egler 37-186.*

9. *Bidens coartata* × *B. molokaiensis*

Putative hybrids of this combination occur on the island of Oahu, the majority of these having been earlier determined as *Bidens macrocarpa* var. *ovatifolia*. The wild hybrids are comparable to vouchers of experimental hybrids of the putative parents (*Gillett 1602 × 1807*). While the *B. molokaiensis* parent is now restricted to Molokai, it is reasonable to postulate its occurrence on Oahu in its evolutionary history. The hybrids are sympatric with the *B. coartata* parent.

OAHU: Oahu, without further locality, *U. S. Expl. Exped. s.n.* (Photograph, Type of *Bidens macrocarpa* var. *ovatifolia* [Gray] Sherff); Kahana Valley, *Degener* et al. *2514* (Type of *B. populifolia* Sherff); Maakue-Papali Ridge, *Cowan 889;* Kaipapau Valley, *Degener* et al. *4114;* Waikane-Schofield trail, *Fosberg 9525;* Puu Kahuauli, *Degener* et al. *19656;* Kaaawa, *Degener* et al. *20875;* Waipio, *Bush and Topping 3780;* Kipapa Gulch, *Hosaka 1107;* Pupakea, *Stone 2795;* Punaluu, *Degener* et al. *10043;* Kalihi-uka, *Caum s.n.*

10. *Bidens coartata* × *B. skottsbergii*

Natural hybrids of this lineage occur on the islands of Lanai and Maui and have been recognized as *Bidens distans* Sherff. The putative hybrids compare remarkably well with experimental hybrids of these species (*Gillett 1602 × 1753*). The hybrids are allopatric with both parents, *B. coartata* being restricted to Oahu and *B. skottsbergii* restricted to Hawaii. Achenes of the latter species and of the hybrids are strongly armed with retrorsely barbed awns, well adapted for dispersal.

LANAI: Mts. near Koele (Gay's) *Forbes 148-L* (Isotype of *Bidens distans* Sherff); Kapano, *Munro s.n.;* Kahinahina, *Munro 505.*

WEST MAUI: Mt. Alani, *Degener 19577;* Iao Needle, *Degener 23650.*

11. *Bidens forbesii* × *B. menziesii*

Hybrids of this lineage occur on Kauai, the principal expression for these centering in the Waimea Canyon area on dry, exposed habitats. Putative hybrid status of this combination is assigned to *Gillett 1888*, used extensively in the crossing program. This collection is from a highly variable natural population in which the foliage character of *Bidens menziesii* is strongly expressed. The hybrids are sympatric with the *B. forbesii* parent.

KAUAI: Nonou Mts., *Forbes 592-K;* Kukui Trail, Waimea Canyon, *Degener et al. 27189;* Laaukahi, Koloa, *St. John and Fosberg 13459;* Kapoahiaola, *Degener et al. 23868;* Waimea drainage basin, west side, *Forbes 811-K* (Type of *Bidens setosa* Sherff); Nawiliwili Bay, *Forbes 704-K* (Type of *Bidens micranthoides* Sherff), *Degener 21530;* Koloa, *Degener 27350;* Kaaweiki Ridge, *Degener 27183;* Kalalau Valley, *Degener 21488.*

12. *Bidens mauiensis* var. *cuneatoides* × *B. molokaiensis*

The natural hybrids are restricted to the slopes and rim of Diamond Head, Oahu and are allopatric with both putative parents. The combination has not been synthesized experimentally, yet the simple, cuneate leaves, the scapose inflorescences, and the broad awnless achenes of the hybrids clearly show the combination of parental genes. The parents occur on Maui and Molokai, respectively.

OAHU: Rim of Diamond Head Crater, *Mrs. G. C. Munro s.n., Earl Ozaki et al. 1464, W. A. Bryan s.n.* (Type of *Bidens cuneata* Sherff).

13. *Bidens fulvescens* × *B. macrocarpa*

The putative parents are restricted to the Waianae Range and the Koolau Range, respectively, on the island of Oahu. It is likely, however, that they have been sympatric in their recent evolutionary history. The hybrids are sympatric with the *Bidens macrocarpa* parent on the Koolau Range, Oahu. The achene characters of the hybrids clearly reflect a combination of the genes of the putative parents. The hybrid combination has not been synthesized, yet it seems clear that in this case nature has succeeded in making the combination. The foliage and inflorescence characters of the parental species and putative hybrids are comparable.

OAHU: Hauula, *Degener et al. 4080* (Type of *Bidens magnidisca* Deg. and Sherff), *Krauss s.n., Degener 11865;* Kalauao-Waimalu Ridge, *St. John 13024;* Kaluanui Valley, *Degener et al. 12284;* Halawa Gulch, *Degener et al. 4133;* Sacred Falls, *Degener et al. 19694;* Mt. Tantalus, *Degener et al. 10066;* east Ridge, Manoa Valley, *Degener et al. 3529.*

14. *Bidens macrocarpa* × *B. wiebkei*

Experimental hybrids of this lineage (*Gillett 1819* × *1603*) compare remarkably well with plants heretofore classified as *Bidens conjuncta* Sherff. Distribution of this material is restricted to West Maui, where material has been collected over a broad range of territory between the Honakahau area, opposite the Molokai habitat of *Bidens wiebkei*, and the eastern slopes of the West Maui volcano. Populations occur on wet uplands on habitats similar to that of *B. macrocarpa* on the Koolau Range of Oahu.

WEST MAUI: Honakahau Drainage, *Forbes 468-M* (Type of *Bidens conjuncta* Sherff); upper Olowalu Valley, *Forbes 2363-M;* Kaanapali, *Rock 8144;* rainy uplands above McGregor, *Degener 22032;* Mt. Eke, *Degener and Wiebke 2164, 2178.*

15. *Bidens menziesii* var. *menziesii* × *B. wiebkei*

The putative hybrids and both parental species are sympatric on the island of Molokai. Experimental plants of the same combination compare well with the cited specimens. The comparison between the experimental F_1 (*Gillett 1819* × *1802*) and the type of *Bidens salicoides* Sherff is remarkably close.

MOLOKAI: Central Molokai, *Uehara s.n.;* Kawela Gulch, *Degener 2894;* Mokomoka Gulch, *Degener 7459; Kalae, Hillebrand s.n.;* East Ohia, *Wiebke 3084* (Type of *Bidens salicoides* Sherff).

LANAI: Waiapoa, *Munro 122.*

16. *Bidens ctenophylla* × *B. menziesii* var. *filiformis*

Gillett made a field study of a hybrid swarm consisting of several hundred plants occurring over an area approximately 4 miles long, extending on old lava flows from 600 to 1,000 meters in elevation, on the Puu Waawaa Ranch, Hawaii. The population includes segregates that closely match the *Bidens menziesii* var. *filiformis* parent in habit achene structure, and leaf form (*Gillett 1745*). The population also includes segregates that closely approach the *B. ctenophylla* parent. Several of the natural hybrids are closely matched by experimental F₁ hybrids (*Gillett 1756 × 1751*). There is in this population an essentially continuous series of intermediates between the parental lines, including perhaps several hundred forms. Most of the hybrids are highly fertile, often bearing a heavy crop of mature achenes. To see this population is to realize the enormous number of segregate types and very great potential of morphological expression through backcrosses and intercrossing with other species. The experimental crosses between the parental species and other species show that such intercrossing is possible. The futility of a taxonomic treatment that would attempt to assign formal descriptions and names to the hybrid segregates is clearly apparent. The obviously great recombination potential in this material, and probably in other hybrid populations of Hawaiian *Bidens* seems to preclude a series of mutually exclusive specific circumscriptions for the various parental lines. Indeed, there would be in this instance a very strong case for reducing both of the parental species to subspecific status.

HAWAII: Kemole, *Rock 8310;* Puu Waawaa lava flows, *Rock 25609;* Puu Waawaa Ranch, above the main cinder cone, 1,000 m., *Gillett 2111-2127;* road west of Puu Waawaa, *Degener* et al. *3814;* Huehue, *Degener* et al. *3813.*

In addition to the first five of the above sixteen putative hybrid combinations, Sherff cited three putative hybrids that we do not accept. These are listed below and the reasons for their rejection are given.

1. *Bidens mauiensis* var. *cuneatoides* × *B. hillebrandiana* (Sherff, 1937)

We have collected and grown material from the region cited as the locality for this hybrid. We have grown the plants (cited here as *Bidens mauiensis* var. *mauiensis* [*Gillett 1872*]) through two generations and have examined carefully the foliage, inflorescences, and mature achenes, finding no evidence of any relationship to *B. hillebrandiana*. The plants produce achenes with broad lateral wings and without terminal awns, achenes that contrast strongly with the narrow, strongly awned achenes of *B. hillebrandiana*. If this material had the slightest relationship to *B. hillebrandiana*, the achenes would have at least vestigial awns, reflecting the epistatic nature of the genes for awns. Although we do not accept Sherff's evidence for this combination, we do cite other material for it (hybrid no. 7 in the above enumeration) that compares well with our experimental vouchers.

2. × *Bidens dimidiata* Deg. and Sherff (Sherff, 1951)

The innumerable unsuccessful attempts we have made to cross *Bidens cosmoides* with other Hawaiian *Bidens* clearly indicate the genetic isolation of this species and the most unlikely possibility of its natural hybridization with any other species. Examination of an isotype (*Degener 20509*) shows it is only a slight variant of *B. cosmoides*, possibly not genotypically different from other populations of the species. The postulation of hybrid status for this collection is in our judgment excessively speculative.

3. *Bidens skottsbergii* × ? (Sherff, 1953)

We are familiar with *Bidens skottsbergii*, having studied it in the field and grown it through three generations. The specimens of this species (*Degener and Nitta 4215, St. John* et al. *11239*) cited by Sherff as of possible hybrid origin have inflorescences, foliage, and achenes that clearly reflect only insignificant, superficial, probably nongenotypic, differences in the expression of *B. skottsbergii*.

TAXONOMIC IMPLICATIONS

Significant taxonomic information brought to light in this study includes evidence of the genetic instability of achene form within the genus *Bidens*. The awnless condition that characterizes the genus *Camplyotheca* is clearly only the expression of relatively few hypostatic genes. We have found that both awned and awnless achenes are produced by the same population (see achenes of *B. wiebkei* in fig. 23, *b, d, f, h*). This information confirms the judgment of Schultz-Bipontinus (1856) and later workers who reduced the genus *Camplyotheca* to taxonomic synonymy under *Bidens*. In addition, it has been shown that achenes with broad lateral wings, a diagnostic character for the genus *Coreopsis*, are evolved within the genus *Bidens* in Hawaii and that the winged expression is regulated by genes that have intermediate dominance (see figs. 6, 9, 18, and fig. 23, *d, g*, showing hybrids between *Bidens mauiensis* [Coreopsis mauiensis A Gray], on the one hand, and *B. hillebrandiana, B. menziesii,* and *B. wiebkei*). This substantiates the judgment of later workers who assigned to *Bidens* certain Polynesian species that Gray (1861) and Drake Del Castillo (1886–1892) had placed under *Coreopsis*.

The expressions of strongly awned achenes, awnless achenes, awnless achenes with broad lateral wings, awnless achenes with helical or longitudinal contortions, and awnless achenes completely enveloped by abaxial receptacular bracts all emphasize the amazing propensity for evolutionary change that characterizes the genus *Bidens* in the Hawaiian Islands. There can be little doubt that these remarkably different expressions have evolved from a common introductory stock. Even *B. cosmoides,* with its distinctive achenes, figure 24, very likely evolved from this vagile stock that probably possessed awned achenes, a necessary device for ingress via long-distance dispersal.

In our review of herbarium material of Hawaiian *Bidens,* a total of ten currently recognized species have been found to be based upon hybrid combinations. These species and their hybrid equivalents are listed below, and it is suggested that the formulae provide a more useful reference to the material by way of indicating probable origins of populations.

Species	Hybrid Combination
Bidens awaluana Deg. and Deg. and Sherff 1964	*Bidens mauiensis* var. *cuneatoides* × *B. menziesii*
B. conjuncta Sherff 1937	*B. macrocarpa* × *B. wiebkei*
B. cuneata Sherff 1937	*B. mauiensis* var. *cuneatoides* × *B. molokaiensis*
B. distans Sherff 1937	*B. coartata* × *B. skottsbergii*
B. graciloides Sherff 1937	*B. hillebrandiana* × *B. wiebkei*
B. magnidisca Sherff 1937	*B. fulvescens* × *B. macrocarpa*
B. micranthoides Sherff 1937	*B. forbesii* × *B. menziesii*
B. populifolia Sherff 1937	*B. coartata* × *B. molokaiensis*
B. salicoides Sherff 1937	*B. menziensii* var. *menziessi* × *B. wiebkei*
B. waianensis Sherff 1937	*B. fulvescens* × *B. menziesii*

If the above ten species were to be deleted from the current number, a reduction of nearly one-fourth of the recognized species would be realized. In our judgment this would confer a much more manageable status on this difficult assemblage of hybrid complexes.

The genus *Bidens* in the Hawaiian Islands is formed of a very large assemblage of taxa that can be delineated into two groups on the basis of biochemical, genetical, and morphological criteria. One of these groups includes a single species, *B. cosmoides*, a most peculiar entity characterized by very large capitula, remarkably elongated styles, and achenes that are enfolded by receptacular bracts. The other group consists of a very broad series of taxa which are closely related on biochemical, genetical, and morphological evidence. If one views the latter group with the criteria of the biological species, he would appraise it as an assemblage of no less than 50, possibly more than 100, intergrading races (not species), the genetical dynamics of which preclude a discrete taxonomic treatment. That these races undergo extensive natural hybridization, generating a vast assemblage of recombinations, is the most significant conclusion of these studies. The obviously close ties between elements of the indigenous Hawaiian *Bidens,* their common chromosome number, and their remarkably diverse habitat relationships clearly indicate that adaptive radiation from a single introduction is the most plausible explanation for the abundance of evolutionary changes that have taken place.

Adaptive radiation has been proposed by Lack (1947) for the remarkable diversity in form of the Helianthioid genus *Scalesia* (Howell, 1941) in the Galapagos Islands. This process, particularly notable in insular habitats, is one of the dominant expressions of evolution in the Hawaiian biota and is apparent in the diversity of the endemic avian family Drepaniidae (Lack, *op. cit.*), also of *Wikstroemia* (Thymelaeaceae), *Cyanea, Clermontia (Lobeliaceae)*, and other genera. It is probably true that the expression of adaptive radiation is more conspicuous and more extensive in the Hawaiian biota than in the Galapagian biota, for the Hawaiian Islands have far greater isolation and greater diversity of soil, topography, and climate than have the Galapagos Islands.

That the current taxonomic treatment of Hawaiian *Bidens* and probably other genera is marked by excessive splitting and scarcely classifiable taxa can hardly be doubted. Existing "keys" to species are grossly unsatisfactory for the reason that the independent segregation of genes, a fundamental law of biology, renders impossible a discrete classification. The extensive hybridization in this material has produced an almost infinite series of recombinations marked by the independent segregation of "key" characters. There is scant possibility that a mutually exclusive, satisfactory "key" will ever be devised for the vast number of recombinations present.

Certainly the description and naming of still more species is unwarranted and will not provide an answer for this problem group. The ultimate result of still further splitting would be the formal recognition of biotypes, suggesting the chaotic disposition accorded such genera as *Rubus, Crataegus,* and *Hieracium.*

DISCUSSION

The present-day populations of Hawaiian *Bidens* undoubtedly evolved under a strong influence of hybridization, associated with the climatic changes and the habitat diversity on these islands. Frequent lava flows and the erosion of precipitous gullies undoubtedly provided for the isolation of small populations conducive to rapid evolution. These autogamous populations could have accomplished a rapid fixation of adaptive gene combinations. The complex variability of Hawaiian *Bidens* may have been the result of the convergence and hybridization of previously isolated populations under the influence of secular climatic changes. Such hybridization could provide for a rapid influx of genes, enhancing the gene pool and extending the range of ecological tolerance. Spatial and ecological barriers, rather than internal genetical barriers, seem to have been most significant in these races. The evolution of awnless achenes, shown by our work to be a comparatively simple process marked by the fixation of hypostatic genes, undoubtedly contributed to the geographical isolation of many races. Ingress upon the habitats of these isolated populations, however, would have been possible by the races with awned achenes.

The fact that the New World has more species of *Bidens* than Polynesia, and the fact that *Bidens* is not known to be present in the indigenous floras of Samoa and Fiji, all point to a New World, (Carlquist, 1965) rather than an Austromalayan (Fosberg, 1948) origin for the Polynesian species of *Bidens*.

Cytological data reveal the close relationships of the Hawaiian *Bidens*. All species and their hybrids are hexaploid with the chromosome number of n = 36. Chromosome pairing at meiosis is perfect in all species and in their hybrids with the exception of two F_1 hybrids where two to four univalents were observed. No multivalents were observed in species or hybrids.

Biochemical data indicate a close relationship among most of these species, for the group affinity comparison values are quite high, with the exception of *Bidens cosmoides.*

The close relationship of these races is shown by the interfertility that prevails over sharp contrasts in habitat adaptation, habit, morphology, and geographical territory. The degree of crossability does not correlate with ecological, geographical, morphological, and biochemical differences, except in the case of *Bidens cosmoides,* a species that is strongly distinct from the other elements on all these criteria, and genetically isolated. These all suggest that *B. cosmoides* has arisen from a very early divergence from the original ingressive stock of *Bidens* in Hawaii. This species clearly is in a category by itself, but its evolutionary history would seem to preclude the sectional designation accorded it by Sherff (1937).

The complete absence of any apparent means of dispersal for the achenes of *B. cosmoides* and its restriction to a limited territory on one island (Kauai) support its status as an early isolate and substantiate the hypothesis developed by Carlquist (1966*b, c*) for the loss of dispersal as an adaptive trend in Pacific Compositae. While the loss of dispersal is conspicuous in the several species of *Bidens* that lack achenial awns and lateral setae, it is most conspicuous in *B. cosmoides* where the mature achenes are almost completely enveloped by abaxial receptacular bracts, with no apparent means of attachment to an agency of dispersal.

Evidence from morphological comparisons of the hybrids with their parents reveal that the following morphological features are relatively epistatic: woody and erect habit; simple leaf form; simple, monocapitate inflorescence, and awned achenes. The variability of F_2 hybrids indicates that the character differences are in the majority of cases regulated by multiple factors.

It is clear from our experimental work that the genes regulating character differences in the leaves and achenes of these species impose relatively broad effects and are relatively few in number (note fig. 15 of leaf types in the experimental F_2 population of *Bidens ctenophylla* × *B. hillebrandiana*). It is also clear that the parental lines are notably heterozygous, for the plants of a given experimental F_1 hybrid population usually are variable and the F_2 hybrids often have conspicuous characters that are not expressed, or are weakly expressed, in the parents.

In his discussion of the five Hawaiian species of *Gouldia* (Rubiaceae), Hillebrand (1888, p. 167) comments:

I am fully aware that as species they are not unimpeachable, that exceptions exist to their definitions, that forms of transition connect most of them, but this is what occurs in the species of all leading Hawaiian genera.

Because of his extensive field studies, access to innumerable pristine habitats prior to their destruction, and his perceptive eye, Hillebrand's judgment and work merit a high order of respect. There can be little doubt that the genus *Bidens* (*Campylotheca* fide Hillebrand) is one of the genera alluded to in Hillebrand's statement. This genus expresses great morphological and ecological diversity and is one of the most frequently encountered of the indigenous genera. The experimental work carried out in this study clearly shows that the prolific evolutionary divergence of *Bidens* has not been accompanied by the genetic isolation of its productions and is marked by extensive natural hybridization. The absence of genetic barriers is so obvious and general as to suggest that there has, indeed, been selection for interspecific compatibility as suggested by Rattenbury (1962).

The wild hybrids of Hawaiian *Bidens* are in most cases found on areas marked by undisturbed vegetation. Changes imposed on these areas are brought about by such natural disturbances as lava flows and the erosion of sea cliffs and canyon walls. Natural hybridization, therefore, has been largely regulated by the accommodation of hybrids by the pristine native habitat and is not related to habitat disturbances imposed by man. Therefore, hybridization is the product of the chance juxtaposition of races through dispersal and migration and the coincidental accommodation of the resultant hybrids by the natural habitat. Under these circumstances, the odds for natural hybridization, even with the absence of internal genetical barriers, are relatively slight. It, therefore, is not surprising that several fertile experimental hybrids have not been seen in nature. A similar discrepancy was noted in New Zealand species of *Epilobium* (Brockie, 1966) where a great many more experimental hybrids were produced than occur in nature.

The relatively common occurrence of hybridization in *Bidens* reflects the general nature of Hawaiian Asteraceae where intergeneric as well as interspecific hybrids are not unusual. Sherff (1935, 1944) and Keck (1936) have shown that the genera *Wilkesia*, *Argyroxiphium*, *Dubautia*, and *Railliardia* are all linked by intermediates in the order given. Keck (*op. cit.*) merged the genus *Wilkesia* into *Argyroxiphium* and placed *Railliardia* under *Dubautia*. Sherff (1944) described an intergeneric hybrid, × *Argyrautia degeneri*, between *Argyroxiphium* (sens. lat.) and *Dubautia* (sens. lat.). While hybridization has been noted for many other families of Hawaiian phanerogams, it probably is most conspicuous in the Asteraceae.

LITERATURE CITED

ALSTON, R. E.
  1967. Biochemical systematics. *In* T. Dobzhansky, M. K. Hecht, and W. C. Steere, eds., Evolutionary biology, 1:197–305. New York: Appleton-Century-Crofts.
ALSTON, R. E., H. RÖSLER, K. NAIFEH, and T. J. MABRY
  1965. Hybrid compounds in natural interspecific hybrids. Proc. Nat. Acad. Sci. U.S.A., 54:1458–1465.
ALSTON, R. E., and J. SIMMONS
  1962. A specific and predictable biochemical anomaly in interspecific hybrids of *Baptisia viridis* × *B. leucantha*. Nature (London). 195:825.
ALSTON, R. E. and B. L. TURNER
  1964. Comparative chemistry of *Baptisia*: problems of interspecific hybridization. *In* C. A. Leone, ed., Taxonomic biochemistry and serology. Pp. 225–238. New York: Ronald Press.
BALGOOY, M. M. J. VAN
  1960. Preliminary plant-geographical analysis of the pacific. Blumea 10:385–430.
BREHM, B. G.
  1963. The distribution of alkaloids, free amino acids, flavonoids, and certain other phenolic compounds in *Baptisia leucophaea* Nutt. var. *laevicaulis* Gray, and their taxonomic implication. Ph.D. dissertation, University of Texas Library, Austin.
BROCKIE, W. B.
  1966. Artificial hybridisation of New Zealand species and varieties of *Epilobium*. New Zealand Jour. Bot. 4:366–391.
BROWN, F. B. H.
  1935. Flora of southeastern Polynesia. III. Dicotyledons. Bishop Mus. Bull., 130:1–386.
CANDOLLE, A. P. DE
  1836. Prodromus. 5:593. Paris: Treuttel et Würtz.
CARLQUIST, S.
  1965. Island Life: A natural history of the islands of the world. Garden City, N.Y.: Natural History Press.
  1966a. The biota of long-distance dispersal. I. Principles of dispersal and evolution. Quart. Rev. Biol., 41:247–270
  1966b. The biota of long-distance disfusal. II. Loss of dispersibility in Pacific Compositae. Evolution, 20:30–48.
  1966c. The biota of long-distance disfusal. III. Loss of dispersibility in the Hawaiian flora. Brittonia, 18:310–335.
CASSINI, H.
  1827. *In* Dictionnaire des Sciences Naturelles. 51:475–476. Paris: Le Normant.
CAVE, M. S., ed.
  1956–1963. Index to plant chromosome numbers, vol. 1 and 2. Chapel Hill: Univ. North Carolina Press.
DEGENER, O.
  1933–1963. Flora Hawaiiensis. Honolulu: Published by the author.
DRAKE DEL CASTILLO, E.
  1886–1892. Illustrations florae insularum maris Pacifici. Paris: G. Masson.
ELLISON, E. L., R. E. ALSTON, and B. L. TURNER
  1962. Methods of presentation of crude biochemical data for systematic purposes, with particular reference to the genus *Bahia* (Compositae). Amer. Jour. Bot., 49:599–604.
FOSBERG, F. R.
  1948. Derivation of the flora of the Hawaiin Islands. *In* E. C. Zimmerman, ed., Insects of Hawaii, 1:107–119. Honolulu: Univ. of Hawaii Press.
GAUDICHAUD, C.
  1826. Le Voyage autour du Monde. Par M. Louis de Freycinet. Bot. 4:464, pl. 85. Paris: Chez Pillet Aîné.

GRAY, A.
  1861. Characters of some Compositae in the collection of the United States South Pacific Exploring Expedition under Captain Wilkes. Proc. Am. Acad. Arts Sci., 5:125–128.
HARNEY, P. M., and W. F. GRANT
  1963. Biochemical anomaly in flower extracts of interspecific hybrids between *Lotus* species. Science, 142:1061.
HILLEBRAND, W.
  1888. Flora of the Hawaiian Islands. Heidelberg: Carl Winter.
HORNE, D. E.
  1965. The distribution of flavonoids in *Baptisia nuttalliana* and *B. lanceolata* and their taxonomic implications. Ph.D. dissertation, University of Texas Library, Austin.
HOWELL, J. T.
  1941. The genus *Scalesia*. Proc. Calif. Acad., 22(11):221–271.
KECK, D. D.
  1936. The Hawaiian Silverswords. Occ. Pap. Bishop Mus. Honolulu, 11(19):1–38.
LACK, D.
  1947. Darwin's Finches. Cambridge: Cambridge Univ. Press.
LESSING, C. F.
  1832. Synopsis generum compositarum. Berlin: Duncher & Hymblot.
LINNAEUS, C.
  1753. Species plantarum. Stockholm.
  1754. Genera plantarum. Ed. 5. Stockholm.
LÖVE, A., and O. T. SOLBRIG
  1964. I.O.P.B. chromosome number reports. I. Taxon, 13:99–110.
MABRY, T. J., J. KAGAN, and H. RÖSLER
  1965. *Baptisia* flavonoids: nuclear magnetic resistance analysis. Phytochemistry, 4:487–493.
MOORE, J. W.
  1933. New and critical plants from Raiatea. Bishop Mus. Bull., 102:46–48.
NECKER, N. J. DE
  1790. Elementa Botanica, 1:86–87.
NUTTALL, T.
  1841. Descriptions of new species and genera of plants. Trans. Am. Philos. Soc., n. Ser., 7:368–369.
ORNDUFF, R., ed.
  1967. Index to plant chromosome numbers for 1965. Regnum Vegetabile 50. Utrecht: Int. Bur. Plant Tax. and Nomencl.
  1968. Index to plant chromosome numbers for 1966. Regnum Vegetabile 55. Ibid.
RATTENBURY, J. A.
  1962. Cyclic hybridization as a survival mechanism in the New Zealand forest flora. Evolution, 16:348–363.
SCHULTZ-BIPONTINUS, C. H.
  1856. Verzeichniss der Cassiniaceen, welche Herr Edelstan Jardin in den Jahren 1853–55 auf den Inseln des stillen Oceans gesammelt hat. Flora, 39(23):353–362.
SHERFF, E. E.
  1935. Revision of *Tetramolopium, Lipochaeta, Dubautia,* and *Railliardia*. Bishop Mus. Bull., 135:1–136.
  1937. The genus *Bidens*. Field Mus. Nat. Hist. Bot. Ser., 16:(I and II).
  1941. Additions to our knowledge of the American and Hawaiian floras. Ibid., 22:434–435.
  1944. Some additions to our knowledge of the flora of the Hawaiian Islands Amer. Jour. Bot., 31:151–161.
  1951a. Miscellaneous notes on new or otherwise noteworthy dicotyledonous plants. Ibid., 38:54–73.
  1951b. Notes upon certain new or otherwise interesting plants of the Hawaiian Islands and Colombia. Chicago Nat. Hist. Mus. Bot. Leaflets no. 3, p. 7.
  1953. Notes on miscellaneous dicotyledonous plants. Ibid., no. 8, p. 25.

1962. Miscellaneous notes on some American and Hawaiian dicotyledons. Occ. Pap. Bishop Mus. Honolulu, 22(12):207–214.

1964. Some recently collected dicotyledonous plants from the Hawaiian Islands and Mexico. Ibid., 23(7):124–126.

SKOTTSBERG, C.

1926. Vascular plants from the Hawaiian Islands. I. Medd. Göteborgs Bot. Trädgärd, 2: 274–275.

1935. Vascular plants from the Hawaiian Islands. II. Ibid., 10:191.

1944. Vascular plants from the Hawaiian Islands. IV. Ibid., 15:507–508.

1953. Chromosome numbers in Hawaiian flowering plants. Preliminary report. Arkiv för Bot., 3:63–70.

SMITH, A. C.

1951. The vegetation and flora of Fiji. Sci. Month., 73:3–15.

STAHL, E.

1965. Thin-layer chromatography. Berlin, New York: Springer-Verlag.

TURNER, B. L., and R. E. ALSTON

1959. Segregation and recombination of chemical constituents in a hybrid swarm of *Baptisia laevicaulis* × *B. viridis* and their taxonomic implications. Amer. Jour. Bot., 46:678–686.

TURNER, B. L., J. H. BEAMAN, and H. F. L. ROCK

1961. Chromosome numbers in the Compositae. V. Mexican and Guatemalan species. Rhodora, 63:121–129.